P9-CQX-837

The Emperor's New Drugs

The Emperor's New Drugs

EXPLODING THE ANTIDEPRESSANT MYTH

IRVING KIRSCH

A MEMBER OF THE PERSEUS BOOKS GROUP
NEW YORK

Published by Basic Books,
A Member of the Perseus Books Group

Published in 2009 by The Random House Group, Ltd. in the UK

Library of Congress Control Number: 2009937860
ISBN: 978-0-465-02016-4

10 9 8 7 6 5 4 3 2 1

For Leo, Alice, and the grandchildren yet to come

'Brahms is the best antidepressant.'

Peter Sproston, 2008

Contents

Brand Names

Generic	American	British
Fluoxetine	Prozac	Prozac
Paroxetine	Paxil	Seroxat
Sertraline	Zoloft	Lustral
Venlafaxine	Effexor	Effexor
Nefazodone	Serzone	Dutonin
Citalopram	Celexa	Cipramil

The information in this book is not a substitute for professional advice on specific emotional issues. Please consult your GP before changing, stopping or starting any medical treatment, specifically antidepressant medication. So far as the author is aware the information given is correct and up to date as at 3 September 2009. The author and publishers disclaim, as far as the law allows, any liability arising directly or indirectly from the use, or misuse, of the information contained in this book.

Acknowledgements

Special thanks are due to Giuliana Mazzoni, David Bassine, Alan Scoboria and Steven Jay Lynn, who carefully read and provided very helpful feedback on a number of chapters. Giuliana, in particular, helped me set the tone of the early chapters. I also thank Joanna Moncrieff, who gently critiqued a rather poorly done first draft of Chapter 4. I hope she likes this version better.

Dan Hind was my first editor at Random House. He approached me with the idea of doing this book after attending a debate in which I participated. His feedback at various stages was exceptionally helpful, as was his confidence and encouragement. He left Random House before the project was finished, but continued to help me with it even after leaving. He was replaced as my editor by Kay Peddle, who was left in the lurch and whom I thank immensely for her substantial help on the final leg of the journey. Mandy Greenfield has been an eagle-eyed copy-editor, and I thank her for catching my oversights.

Thanks are also due to numerous colleagues and friends who provided helpful comments and information beyond that which I had found in books and journal articles. These include David Antonuccio, David Burns, David Goldberg, David Healy, Steven Hollon, Ted Kaptchuck, Peter Lewinsohn, John and Madge Manfred, Helen Mayberg, Lee Park, Forrest Scogin, Harriet Vickery, Tor Wager and Nelda Wray.

Finally, I thank the wonderful scientists with whom I have collaborated on the research leading to this book: Guy Sapirstein, Thomas Moore, Alan Scoboria, Blair Johnson, Brett Deacon, Tania Huedo-Medina, Joanna Moncrieff, Corrado Barbui, Andrea Cipriani, Sarah Nicholls and David Antonuccio. Research is a team effort, and these colleagues have made wonderful teams.

Preface

Like most people, I used to think that antidepressants worked. As a clinical psychologist, I referred depressed psychotherapy clients to psychiatric colleagues for the prescription of medication, believing that it might help. Sometimes the antidepressant seemed to work; sometimes it did not. When it did work, I assumed it was the active ingredient in the antidepressant that was helping my clients cope with their psychological condition.

According to drug companies, more than 80 per cent of depressed patients can be treated successfully by antidepressants. Claims like this made these medications one of the most widely prescribed class of prescription drugs in the world, with global sales that make it a $19-billion-a-year industry.[1] Newspaper and magazine articles heralded antidepressants as miracle drugs that had changed the lives of millions of people. Depression, we were told, is an illness – a disease of the brain that can be cured by medication. I was not so sure that depression was really an illness, but I did believe that the drugs worked and that they could be a helpful adjunct to psychotherapy for very severely depressed clients. That is why I referred these clients to psychiatrists who could prescribe antidepressants that the clients could take while continuing in psychotherapy to work on the psychological issues that had made them depressed.

But was it really the drug they were taking that made my clients feel better? Perhaps I should have suspected that the improvement they reported might not have been a drug effect. People obtain considerable benefits from many medications, but they also can experience symptom improvement just by knowing they are being treated. This is called the placebo effect. As a researcher at the University of Connecticut, I had been studying placebo effects for many years. I was well aware of the power of belief to alleviate depression, and I understood that this was an important part of any treatment, be it psychological or pharmacological. But I also believed that antidepressant drugs added something substantial over and beyond the placebo effect. As I wrote in my first book, 'comparisons of anti-depressive medication with placebo pills indicate that the former has a greater effect . . . the existing data suggest a pharmacologically specific effect of imipramine on depression'. As a researcher, I trusted the data as it had been presented in the published literature. I believed that antidepressants like imipramine were highly effective drugs, and I referred to this as 'the established superiority of imipramine over placebo treatment'.[2]

When I began the research that I describe in this book, I was not particularly interested in investigating the effects of antidepressants. But I was definitely interested in investigating placebo effects wherever I could find them, and it seemed to me that depression was a perfect place to look. Why did I expect to find a large placebo effect in the treatment of depression? If you ask depressed people to tell you what the most depressing thing in their lives is, many answer that it is their depression. Clinical depression is a debilitating condition. People with severe depression feel unbearably sad and anxious, at times to the point of considering suicide as a way to relieve the burden. They may be racked with feelings of worthlessness and guilt. Many suffer from insomnia, whereas others sleep too much and find it difficult to get out of bed in the morning. Some have difficulty concentrating and have lost interest in all of the activities that previously brought pleasure and meaning into their lives. Worst of all, they feel hopeless about ever recovering from this terrible state, and this sense of hopelessness may lead them to feel that life is not worth living.

In short, depression is depressing. John Teasdale, a leading researcher on depression at Oxford and Cambridge universities, labelled this phenomenon 'depression about depression' and claimed that effective treatments for depression work – at least in part – by altering the sense of hopelessness that comes from being depressed about one's own depression.[3]

Whereas hopelessness is a central feature of depression, hope lies at the core of the placebo effect. Placebos instil hope in patients by promising them relief from their distress. Genuine medical treatments also instil hope, and this is the placebo component of their effectiveness. When the promise of relief instils hope, it counters a fundamental attribute of depression. Indeed, it is difficult to imagine any treatment successfully treating depression without reducing the sense of hopelessness that depressed people feel. Conversely, any treatment that reduces hopelessness must also assuage depression. So a convincing placebo ought to relieve depression.

It was with that in mind that one of my postgraduate students, Guy Sapirstein, and I set out to investigate the placebo effect in depression – an investigation that I describe in the first chapter of this book, and that produced the first of a series of surprises that transformed my views about antidepressants and their role in the treatment of depression.[4] In this book I invite you to share this journey in which I moved from acceptance to dissent, and finally to a thorough rejection of the conventional view of antidepressants.

The drug companies claimed – and still maintain – that the effectiveness of antidepressants has been proven in published clinical trials showing that the drugs are substantially better than placebos (dummy pills with no active ingredients at all). But the data that Sapirstein and I examined told a very different story. Although many depressed patients improve when given medication, so do many who are given a placebo, and the difference between the drug response and the placebo response is not all that great. What the published studies really indicate is that most of the improvement shown by depressed people when they take antidepressants is due to the placebo effect.

Our finding that most of the effects of antidepressants could be explained as a placebo effect was only the first of a number of surprises that changed my views about antidepressants. Following up on this research, I learned that the published clinical trials we had analysed were not the only studies assessing the effectiveness of antidepressants. I discovered that approximately 40 per cent of the clinical trials conducted had been withheld from publication by the drug companies that had sponsored them. By and large, these were studies that had failed to show a significant benefit from taking the actual drug. When we analysed all of the data – those that had been published and those that had been suppressed – my colleagues and I were led to the inescapable conclusion that antidepressants are little more than active placebos, drugs with very little specific therapeutic benefit, but with serious side effects. I describe these analyses – and the reaction to them – in Chapters 3 and 4.

How can this be? Before a new drug is put on the market, it is subjected to rigorous testing. The drug companies sponsor expensive clinical trials, in which some patients are given medication and others are given placebos. The drug is considered effective only if patients given the real drug improve significantly more than patients given the placebos. Reports of these trials are then sent out to medical journals, where they are subjected to rigorous peer review before they are published. They are also sent to regulatory agencies, like the Food and Drug Administration (FDA) in the US, the Medicines and Healthcare products Regulatory Agency (MHRA) in the UK and the European Medicine Agency (EMEA) in the EU. These regulatory agencies carefully review the data on safety and effectiveness, before deciding whether to approve the drugs for marketing. So there must be substantial evidence backing the effectiveness of any medication that has reached the market.

And yet I remain convinced that antidepressant drugs are not effective treatments and that the idea of depression as a chemical imbalance in the brain is a myth. When I began to write this book, my claim was more modest. I believed that the clinical effectiveness of antidepressants had not been proven for most of the millions of patients to whom they are prescribed, but I also acknowledged that they might be beneficial to at least a subset of depressed patients. During the process of putting all of the data together, those that I had analysed over the years and newer data that have just recently seen the light of day, I realized that the situation was even worse than I thought. The belief that antidepressants can cure depression chemically is simply wrong.

In this book I will share with you the process by which I came to this conclusion and the scientific evidence on which it is based. This includes evidence that was known to the pharmaceutical companies and to regulatory agencies, but that was intentionally withheld from prescribing physicians, their patients and even from the National Institute for Health and Clinical Excellence (NICE) when it was drawing up treatment guidelines for the National Health Service (NHS) in the UK.

My colleagues and I obtained some of these hidden data by using the Freedom of Information Act in the US. We analysed the data and submitted the results for peer review to medical and psychological journals, where they were then published.[5] Our analyses have become the focus of a national and international debate, in which many doctors have changed their prescribing habits and others have reacted with anger and incredulity. My intention in this book is to present the data in a plain and straightforward way, so that you will be able to decide for yourself whether my conclusions about antidepressants are justified.

The conventional view of depression is that it is caused by a chemical imbalance in the brain. The basis for this idea was the belief that antidepressant drugs were effective treatments. Our analyses showing that most – if not all – of the effects of these medications are really placebo effects challenges this widespread view of depression. In Chapter 4 I examine the chemical-imbalance theory. You may be surprised to learn that it is actually a rather controversial theory and that there is not much scientific evidence to support it. While writing this chapter I came to an even stronger conclusion. It is not just that there is not much supportive evidence; rather, there is a ton of data indicating that the chemical-imbalance theory is simply wrong.

The chemical effect of antidepressant drugs may be small or even non-existent, but these medications do produce a powerful placebo effect. In Chapters 5 and 6 I examine the placebo effect itself. I look at the myriad of effects that placebos have been shown to have and explore the theories of how these effects are produced. I explain how placebos are able to produce substantial relief from depression, almost as much as that produced by medication, and the implications that this has for the treatment of depression.

Finally, in Chapter 7, I describe some of the alternatives to medication for the treatment of depression and assess the evidence for their effectiveness. One of my aims is to provide essential scientifically grounded information for making informed choices between the various treatment options that are available.

Much of what I write in this book will seem controversial, but it is all thoroughly grounded on scientific evidence – evidence that I describe in detail in this book. Furthermore, as controversial as my conclusions seem, there has been a growing acceptance of them. NICE has acknowledged the failure of antidepressant treatment to provide clinically meaningful benefits to most depressed patients; the UK government has instituted plans for providing alternative treatments; and neuroscientists have noted the inability of the chemical-imbalance theory to explain depression.[6] We seem to be on the cusp of a revolution in the way we understand and treat depression.

Learning the facts behind the myths about antidepressants has been, for me, a journey of discovery. It was a journey filled with shocks and surprises – surprises about how drugs are tested and how they are approved, what doctors are told and what is kept hidden from them, what regulatory agencies know and what they don't want you to know, and the myth of depression as a brain disease. I would like to share that journey with you. Perhaps you will find it as surprising and shocking as I did. It is my hope that making this information public will foster changes in the way new drugs are tested and approved in the future, in the public availability of the data and in the treatment of depression.

1

Listening to Prozac, but Hearing Placebo

In 1995 Guy Sapirstein and I set out to assess the placebo effect in the treatment of depression. Instead of doing a brand-new study, we decided to pool the results of previous studies in which placebos had been used to treat depression and analyse them together. What we did is called a meta-analysis, and it is a common technique for making sense of the data when a large number of studies have been done to answer a particular question. It was once considered somewhat controversial, but meta-analyses are now common features in all of the leading medical journals. Indeed, it is hard to see how one could interpret the results of large numbers of studies without the aid of a meta-analysis.

In doing our meta-analysis, it was not enough to find studies in which depressed patients had been given placebos. We also needed to find studies in which depression had been tracked in patients who were not given any treatment at all. This was to make sure that any effect we found was really due to the administration of the placebo. To better understand the reason for this, imagine that you are investigating a new remedy for colds. If the patients are given the new medicine, they get better. If they are given placebos, they also get better. Seeing these data, you might be tempted to think that the improvement was a placebo effect.

But people recover from colds even if you give them nothing at all. So when the patients in our imaginary study took a dummy pill and their colds got better, the improvement may have had nothing to do with the placebo effect. It might simply have been due to the passage of time and the fact that colds are short-lasting illnesses.

Spontaneous improvement is not limited to colds. It can also happen when people are depressed. Because people sometimes recover from bouts of depression with no treatment at all, seeing that a person has become less depressed after taking a placebo does not mean that the person has experienced a placebo effect. The improvement could have been due to any of a number of other factors. For example, people can get better because of positive changes in life circumstances, such as finding a job after a period of unemployment or meeting a new romantic partner. Improvement can also be facilitated by the loving support of friends and family. Sometimes a good friend can function as a surrogate therapist. In fact, a very influential book on psychotherapy bore the title *Psychotherapy: The Purchase of Friendship*.[1] The author did not claim that psychotherapy was merely friendship, but the title does make the point that it can be very therapeutic to have a friend who is empathic and knows how to listen.

The point is that without comparing the effect of placebos against rates of spontaneous recovery, it is impossible to assess the placebo effect. Just as we have to control for the placebo effect to evaluate the effect of a drug, so too we have to control for the passage of time when assessing the placebo effect. The drug effect is the difference between what happens when people are given the active drug and what happens when they are given the placebo. Analogously, the placebo effect is the difference between what happens when people are given placebos and what happens when they are not treated at all.

It is rare for a study to focus on the placebo effect – or on the effect of the simple passage of time, for that matter. So where were we to find our placebo data and no-treatment data? We found our placebo data in clinical studies of antidepressants, and our no-treatment data in clinical studies of psychotherapy. It is common to have no-treatment or wait-list control groups in studies of the effects of psychotherapy. These groups consist of patients who are not given any treatment at all during the course of the study, although they may be placed on a wait list and given treatment after the research is concluded.

For the purpose of our research, Sapirstein and I were not particularly interested in the effects of the antidepressants or psychotherapy. What we were interested in was the placebo effect. But since we had the treatment data to hand, we looked at them as well. And, as it turned out, it was the comparison of drug and placebo that proved to be the most interesting part of our study.

All told, we analysed 38 clinical trials involving more than 3,000 depressed patients. We looked at the average improvement during the course of the study in each of the four types of groups: drug, placebo, psychotherapy and no-treatment. I am going to use a graph here (Figure 1.1, overleaf) to show what the data tell us. Although the text will have a couple more such charts, I am going to keep them to a minimum. But this is one that I think we need, to make the point clearly. What the graph shows is that there was substantial improvement in both the drug and psychotherapy groups. People got better when given either form of treatment, and the difference between the two was not significant. People also got better when given placebos, and here too the improvement was remarkably large, although not as great as the improvement following drugs or psychotherapy. In contrast, the patients who had not been given any treatment at all showed relatively little improvement.

The first thing to notice in this graph is the difference in improvement between patients given placebos and patients not given any treatment at all. This difference shows that most of the improvement in the placebo groups was produced by the fact that they had been given placebos. The reduction in depression that people experienced was not just caused by the passage of time, the natural course of depression or any of the other factors that might produce an improvement in untreated patients. It was a placebo effect, and it was powerful.

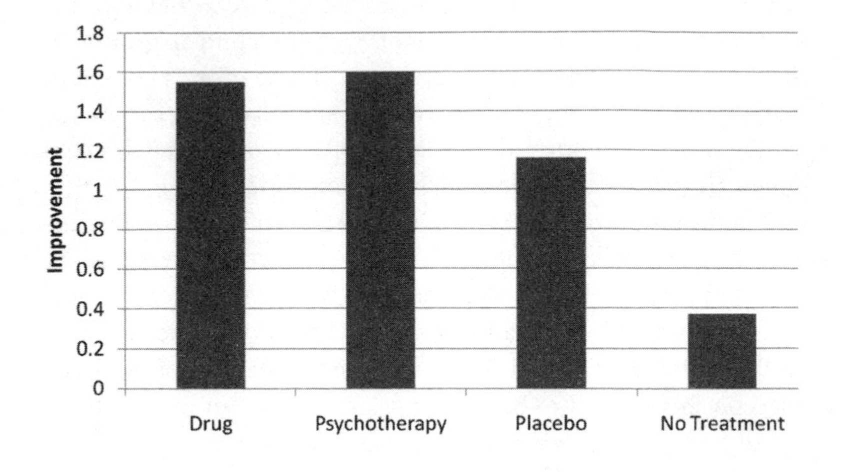

Figure 1.1. Average improvement on drug, psychotherapy, placebo and no treatment.[2] 'Improvement' refers to the reduction of symptoms on scales used to measure depression. The numbers are called 'effect sizes'. They are commonly used when the results of different studies are pooled together. Typically, effect sizes of 0.5 are considered moderate, whereas effect sizes of 0.8 are considered large. So the graph shows that antidepressants, psychotherapy and placebos produce large changes in the symptoms of depression, but there was only a relatively small average improvement in people who were not given any treatment at all.

One thing to learn from these data is that doing nothing is not the best way to respond to depression. People should not just wait to recover spontaneously from clinical depression, nor should they be expected just to snap out of it. There may be some improvement that is associated with the simple passage of time, but compared to doing nothing at all, treatment – even if it is just placebo treatment – provides substantial benefit.

Sapirstein and I were not surprised to find that there was a powerful placebo effect in the treatment of depression. Actually, we were quite pleased. That was our hypothesis and our reason for doing the study. What did surprise us, however, was how small the difference was between the response to the drug and the response to the placebo. That difference is the drug effect. Although the drug effect in the published clinical trials that we had analysed was statistically significant, it was much smaller than we had anticipated. Much of the therapeutic response to the drug was due to the placebo effect. The relatively small size of the drug effect was the first of a series of surprises that the anti-depressant data had in store for us.

One way to understand the size of the drug effect is to think about it as only a part of the improvement that patients experience when taking medication. Part of the improvement might be spontaneous – that is, it might have occurred without any treatment at all – and part may be a placebo effect. What is left over after you subtract spontaneous improvement and the placebo effect is the drug effect. You can see in Figure 1.1 that improvement in patients who had been given a placebo was about 75 per cent of the response to the real medication. That means that only 25 per cent of the benefit of antidepressant treatment was really due to the chemical effect of the drug. It also means that 50 per cent of the improvement was a placebo effect. In other words, the placebo effect was twice as large as the drug effect.

The drug effect seemed rather small to us, considering that these medications had been heralded as a revolution in the treatment of depression – blockbuster drugs that have been prescribed to hundreds of millions of patients, with annual sales totalling billions of pounds.[3] Sapirstein and I must have done something wrong in either collecting or analysing the data. But what? We spent months trying to figure it out.

ARE ALL DRUGS CREATED EQUAL? DOUBLE-BLIND OR DOUBLE-TALK

One thing that occurred to us, when considering how surprisingly small the drug effect was in the clinical trials we had analysed, was that a number of different medications had been assessed in those studies. Perhaps some of them were effective, whereas others were not. If this were the case, we had underestimated the benefits of effective drugs by lumping them together with ineffective medications. So before we sent our paper out for review, we went back to the data and examined the types of drugs that had been administered in each of the clinical trials in our meta-analysis.

We found that some of these trials had assessed tricyclic antidepressants, an older type of medication that was the most commonly used antidepressant in the 1960s and 1970s. In other trials, the focus was on selective serotonin reuptake inhibitors (SSRIs) like Prozac (fluoxetine), the first of the 'new-generation' drugs that replaced tricyclics as the top-selling type of antidepressant. And there were other types of antidepressants investigated in these trials as well. When we reanalysed the data, examining the drug effect and the placebo effect for each type of medication separately, we found that the diversity of drugs had not affected the outcome of our analysis. In fact, the data were remarkably consistent – much more so than is usually the case when one analyses different groups of data. Not only did all of these medications produce the same degree of improvement in depression, but also, in each case, only 25 per cent of the improvement was due to the effect of the drug. The rest could be explained by the passage of time and the placebo effect.

The lack of difference we found between one class of antidepressants and another is now a rather frequent finding in antidepressant research.[4] The newer antidepressants (SSRIs, for example) are no more effective than the older medications. Their advantage is that their side effects are less troubling, so that patients are more likely to stay on them rather than discontinue treatment. Still, the consistency of the size of the drug effect was surprising. It was not just that the percentages were close; they were virtually identical. They ranged from 24 to 26 per cent. At the time I thought, 'What a nice coincidence! It will look great in a PowerPoint slide when I am invited to speak on this topic.' But since then I have been struck by similar instances in which the consistency of the data is remarkable, and it is part of what has transformed me from a doubter to a disbeliever. I will note similar consistencies as we encounter them in this book.

The consistency of the effects of different types of antidepressants meant that we had not underestimated the antidepressant drug effect by lumping together the effects of more effective and less effective drugs. But our re-examination of the data in our meta-analysis held another surprise for us. Some of the medications we had analysed were not antidepressants at all, even though they had been evaluated for their effects on depression. One was a barbiturate – a depressant that had been used as a sleeping aid, before being replaced by less dangerous medications. Another was a benzodiazepine – a sedative that has largely replaced the more dangerous barbiturates. Yet another was a synthetic thyroid hormone that had been given to depressed patients who did not have a thyroid disorder. Although none of these drugs are considered antidepressants, their effects on depression were every bit as great as those of antidepressants and significantly better than placebos. Joanna Moncrieff, a psychiatrist at University College London, has since listed other drugs that have been shown to be as effective as medications for depression.[5] These include antipsychotic drugs, stimulants and herbal remedies. Opiates are also better than placebos, but I have not seen them compared to antidepressants.

If sedatives, barbiturates, antipsychotic drugs, stimulants, opiates and thyroid medications all outperform inert placebos in the treatment of depression, does this mean that any active drug can function as an antidepressant? Apparently not. In September 1998 the pharmaceutical company Merck announced the discovery of a novel antidepressant with a completely different mode of action than other medications for depression. This new drug, which they later marketed under the trade name Emend for the prevention of nausea and vomiting due to chemotherapy, seemed to show considerable promise as an antidepressant in early clinical trials. Four months later the company announced its decision to pull the plug on the drug as a treatment for depression. The reason? It could not find a significant benefit for the active drug over placebos in subsequent clinical trials. This was unfortunate for a number of reasons. One is that the announcement caused a 5 per cent drop in the value of the company's stock. Another is that the drug had an important advantage over current antidepressants – it produced substantially fewer side effects. The relative lack of side effects had been one reason for the enthusiasm about Merck's new antidepressant. However, it may also have been the reason for its subsequent failure in controlled clinical trials. It seems that easily noticeable side effects are needed to show antidepressant benefit for an active drug compared to a placebo.[6]

At first, Sapirstein and I found the equivalence between antidepressants and other drugs puzzling, to say the least. Why should drugs that are not antidepressants be as effective as antidepressants in treating depression? To answer this question, we asked another. What do all these diverse drugs have in common that they do not share with inert placebos? What do SSRIs have in common with the older tricyclic antidepressants, with other less common antidepressants, and even with tranquillizers, depressants and thyroid medication? The only common factor that we were able to note was that they all produce easily noticeable side effects – the one thing that was lacking in Merck's new treatment for depression. Placebos can also produce side effects, but they do so to a much lesser extent than active medication. Clinical trials show that whereas the therapeutic benefits of antidepressants are relatively small when compared to placebos, the difference in side effects is substantial.[7]

Why are side effects important? Imagine that you have been recruited for a clinical trial of an antidepressant medication. As part of the required informed-consent procedure, you are told that you may be given a placebo instead of the active medication, but because this is a double-blind trial, you will not be told which you are getting until the study is over. You are told that it may take weeks before the therapeutic effects of the drug are apparent, and also that the drug has been reported to produce side effects in some patients. Furthermore, as required by the informed-consent procedures that need to be followed in clinical trials, you are also told exactly what those side effects are (for example, a dry mouth, drowsiness, diarrhoea, nausea, forgetfulness) and that these are most likely to occur soon after treatment has begun – before the therapeutic effects are felt.[8]

Now if I were a patient in one of these trials, I would wonder to which condition I had been assigned. Had I been put in the active-drug group or in the placebo group? Hmm, my mouth is getting dry, and I'm beginning to feel a little nauseous. Normally, I might feel distressed by these symptoms, but I have been informed that these are side effects of the active drug. So instead of feeling distressed, I am elated. My dry mouth and nauseous stomach tell me that I have been given the active drug, rather than the placebo. I'm starting to feel better already.

Figuring out whether you have been given the drug or the placebo in a clinical trial is referred to as 'breaking blind'. Clinical trials are supposed to be double-blind. This means that neither the patient nor the doctor is supposed to know whether the patient has been given the active drug or the placebo. In fact, these trials are not really double-blind. Many of the patients break blind, and so do the physicians who are treating them. Both the patients and their doctors come to realize which condition they are in, before being told at the end of the trial. We know this from antidepressant studies in which patients and doctors are asked to say whether they have been given drug or placebo. If they were only guessing, they should be right about half the time, but in fact they are much more accurate than that. In the largest study of this type, 80 per cent of patients accurately identified whether they were on drug or placebo, and in 87 per cent of the cases their doctors also guessed correctly. With the number of patients assessed in this study, the odds of 80 per cent guessing correctly just by chance is less than one in a million. This means that most patients and most doctors broke blind.

For patients, this was especially true if they were in the real drug condition: 89 per cent of patients given the real antidepressants correctly figured out that they were in the drug group. In contrast, only 59 per cent of patients in the placebo group guessed correctly.[9]

ANTIDEPRESSANTS AS ACTIVE PLACEBOS

The breaking of blind by patients in clinical trials may be the key to understanding why all types of different drugs in our meta-analysis, even those that were not antidepressants, had the same effect on depression. When patients are kept blind, they do not know whether they have been given the drug or the placebo. Hence, their expectation of getting better is tempered by their knowledge that they might have been given a placebo. But when they break blind, their expectations change. If they know they have been given the active drug, rather than the placebo, they become much more confident of improving. Conversely, if they realize that they are in the placebo group, their expectancy of improvement declines.

As we shall see in Chapter 6, expectations of improvement are a central factor in the placebo effect. People expect to get better when given a treatment, and in many conditions that expectation can produce the improvement they expect as a sort of self-fulfilling prophecy. In other words, patients who break blind in clinical trials might improve more on the active drug than on the placebo, simply because they know they are getting a real drug rather than a sugar pill. If they believe they are on the active drug, they have a greater expectation of improvement, and because of these enhanced expectations they actually do improve more. On the other hand, if they realize they have been given a placebo, they expect – and therefore experience – less improvement.

This is not just speculation. It is backed by evidence. Some antidepressant trials are conducted without placebo groups.

These are called comparator trials, because they compare one antidepressant to another. In comparator trials, all patients are given an active drug, and they know that there is no chance at all of getting a placebo. A group of researchers led by Joel Sneed at Columbia University in New York compared the response of patients in comparator trials to that of patients in placebo-controlled trials. The researchers found that patients in the comparator trials were significantly more likely to improve. Specifically, 60 per cent of patients responded to antidepressants in the comparator trials, but only 46 per cent were rated as improved in the placebo-controlled trials.[10] This difference resulted from patients knowing that they were definitely getting an active drug versus knowing that they might be getting a placebo, as that was the only difference between the two types of trials that were compared. Because it was produced by what the patients believed about the drug, rather than by the drug itself, it can be considered a placebo effect.

To summarize the argument to this point, we found a relatively small difference between the response to antidepressant drugs and the response to placebos. In other words, the drug effect was rather small. We also found that the small but significant difference between active drugs and placebos was not limited to antidepressants. Other active drugs also reduced depression more than placebos did. The one thing that all of these drugs had in common was that they produced side effects, and side effects have been associated with figuring out whether one has been given an active drug or a placebo in a clinical trial. Finally, we have seen that knowing that one is getting an active drug boosts the effectiveness of the drug, and knowing that one might have been given a placebo decreases its effectiveness. Putting all of this together leads to the conclusion that the relatively small difference between drugs and placebos might not be a real drug effect at all. Instead, it might be an enhanced placebo effect, produced by the fact that some patients have broken blind and have come to realize whether they were given drug or placebo. If this is the case, then there is no real antidepressant drug effect at all. Rather than comparing placebo to drug, we have been comparing 'regular' placebos to 'extra-strength' placebos.

When Sapirstein and I published our analysis, we could not prove that the difference between active drug and placebo in antidepressant trials was due to an enhanced placebo effect. Given the data that we had, this was only a hypothesis, but it was a hypothesis based on substantial circumstantial evidence. Besides the data I summarized in the last paragraph, there are two additional kinds of evidence that support the enhanced placebo effect hypothesis. One of these is that there is an exceptionally high correlation between improvement and the experience of SSRI side effects.[11] One might expect to find a negative association between side effects and improvement. Side effects of SSRIs include sexual dysfunction, insomnia, short-term weight loss, long-term weight gain, diarrhoea, nausea, drowsiness, skin reactions, nervousness, anorexia, dry mouth and sweating.[12] One would think that experiences like this would make people feel more depressed. Indeed, some of these side effects could also be interpreted as symptoms of depression. But in fact the relationship is in the opposite direction. The more side effects a person experiences when taking Prozac, the more he or she improves on the drug. I can think of only one reason why insomnia, diarrhoea and nausea might be linked to improvement, and that is that they lead patients to conclude that they have been given the active drug, rather than the placebo.

The association between side effects and improvement is so strong as to be almost perfect. Correlations can range from zero to one. The correlation between side effects and improvement when taking Prozac is .96, which is just about as high as a correlation can get.[13] It is exceptionally rare to find correlations this high in research. My colleague John Kihlstrom at the University of California at Berkeley calls data like this 'Faustian' – by which he means that researchers would sell their souls to obtain them.[14] A high correlation between two things does not mean that one has caused the other. Hat sizes and shoe sizes are highly correlated, but big feet do not cause swollen heads.

Similarly, the correlation between side effects and improvement does not prove that side effects produce the improvement. Still, it fits the enhanced placebo hypothesis perfectly, and it is hard to think of another explanation for it.

While writing this book, I was invited to speak about my antidepressant research by Corrado Barbui and Andrea Cipriani, psychiatrists at the University of Verona who had conducted studies with results similar to mine, but who still believed that antidepressants had a chemical effect.[15] After my talk, we argued a bit about my contention that the small differences between antidepressant drugs and placebos might be due to the presence of side effects and the consequent breaking of blind among patients who had been given the real drug rather than the placebo. 'If you are right about that,' said these two gentlemen of Verona, 'then controlling for side effects statistically ought to eliminate the drug effect completely.' I agreed, and we decided to test this hypothesis using their collection of all the published and unpublished clinical trials that GlaxoSmithKline had conducted on their SSRI, Seroxat. The results of that analysis showed that once you adjust for drug–placebo differences in side effects, differences in rates of improvement are no longer statistically significant.[16]

Another kind of evidence supporting the active placebo hypothesis comes from studies comparing antidepressants to what are called 'active placebos'. An active placebo is a real drug that produces side effects, but that should not have any therapeutic benefits for the condition being treated. It is used to prevent patients in clinical trials from breaking blind – that is, from guessing the condition to which they have been assigned on the basis of side effects. If the experience of side effects leads patients to conclude that they are in the drug group, rather than the placebo group, then the use of active placebos should keep them in the dark.

What would happen if active placebos were used in clinical trials, rather than inactive placebos? Would one still get the relatively small but significant difference between drug and placebo? We already have the beginning of an answer to this question.

Active placebos have been compared to antidepressants in nine clinical trials.[17] In these trials, the drug atropine was used as an active placebo. Atropine is an active medication. It is used in the treatment of gastric dysfunctions such as irritable bowel syndrome, diarrhoea and peptic ulcers. It can also be used to treat motion sickness, bed-wetting and symptoms of Parkinson's disease, but it is not an antidepressant. Its side effects include a dry mouth, insomnia, headaches and drowsiness, which have also been reported as side effects of antidepressants. It has significantly fewer side effects than the antidepressants to which it was compared in these trials, but should still help prevent patients from breaking blind and realizing that they have been given a placebo, at least to some degree.

Most of the published clinical trials comparing antidepressants to inert placebos – that is, placebos that do not produce side effects – show significant differences between the active drug and the placebo. When an active placebo is used, most clinical trials do not show a significant benefit for antidepressants. Of the nine clinical trials in which an antidepressant was compared to atropine, a significant difference between drug and placebo was found in only two. Furthermore, in the two studies that asked raters to guess which patients had been given antidepressants and which had been given the active placebo, the raters were able to guess what medication had been given at better-than-chance levels. Despite this, in the vast majority (78 per cent) of the clinical trials in which active placebos were used, no significant differences were found between the drug and the placebo. So comparisons with inactive placebos are much more likely to show drug–placebo differences than comparisons with active placebos. This suggests that at least part of the difference that has been found between antidepressant and placebo may be due to the experience of more side effects on the active drug than on the placebo.

Let's summarize the arguments for the active placebo hypothesis.

1. Antidepressants produce significantly more side effects than inert placebos.
2. Most patients in clinical trials are able to figure out whether they have been assigned to the drug group or the placebo group before being told.
3. There are relatively small but significant differences between active drugs and inert placebos, and these differences are independent of the type of active drug that is used. Indeed, the active drug need not even be an antidepressant.
4. Although a drug need not be an antidepressant to be more effective than a placebo, it does seem to need sufficient side effects that patients can figure out that they have not been given a placebo.
5. When antidepressants are compared to active placebos, differences in outcome are substantially harder to find.
6. The more side effects that depressed patients experience on the active drug, the more they improve.
7. When you control for differences in side effects, drug–placebo differences in improvement are not statistically significant.

Taken together, these data strongly support the idea that side effects lead clinical-trial patients to realize they have been given the active drug, and that this realization leads them to improve more than patients in the placebo groups. It may not be conclusive proof, but it is strong evidence.

* * *

In this chapter we have looked at the results of published clinical trials of antidepressant medication. The published studies showed a significant, but surprisingly small, effect of antidepressants over placebos. But as I noted at the beginning of the chapter, those data represented only the beginning. As I later discovered, there were also studies that had been withheld from publication. These unpublished studies were clinical trials that did not show a significant benefit for drugs over placebo medication – trials that the drug companies withheld from public scrutiny. In the next chapter I describe the process by which I learned about the hidden clinical trials and how not only the drug companies, but also regulatory agencies, kept the data from the public.

2

The 'Dirty Little Secret'

When we wrote up our meta-analysis for publication, Sapirstein and I were cautious in our interpretation of the data. Despite our concerns about patients breaking blind and realizing whether they were in the drug group or the placebo group, we concluded that our results showed 'a considerable benefit of medication over placebo'. Nevertheless, the article reporting our analysis of the published literature proved to be highly controversial – controversial enough for the editors of the journal to insert a warning label at the beginning, much like the warning label that you find on packs of cigarettes or, more recently, on patient information leaflets for antidepressants. They wrote:

> The article that follows is a controversial one. It reaches a controversial conclusion – that much of the therapeutic benefit of antidepressant medications actually derives from placebo responding. The article reaches this conclusion by utilizing a controversial statistical approach – meta-analysis. And it employs meta-analysis controversially – by meta-analysing studies that are very heterogeneous in subject selection criteria, treatments employed, and statistical methods used. Nonetheless, we have chosen to publish the article. We have done so because a number of the colleagues who originally reviewed the manuscript believed it had considerable merit, even while they recognized the clearly contentious conclusions it

> reached and the clearly arguable statistical methods it employed. The article that follows is a controversial one. It reaches a controversial conclusion – that much of the therapeutic benefit of antidepressant medications actually derives from placebo responding.[1]

In the decade that has passed since our article was published, the dust has settled around the issue of meta-analysis. It is no longer considered a controversial procedure. Meta-analyses of clinical trials are now routinely published in all of the top medical journals, and the National Institute for Health and Clinical Excellence (NICE), which publishes the treatment guidelines that are used by the NHS, crafts recommendations on the basis of meta-analyses that it conducts. Nevertheless, the editors were right about our article being controversial. Although some scholars in the field were persuaded by our analyses, others were sceptical, to put it mildly.[2] The sceptics knew that antidepressants worked – if we had found otherwise, we must have done something wrong. Certainly there were other clinical trials of antidepressants beyond those that we had included in our analyses. Surely an analysis of those studies would point to a different conclusion.

There were indeed clinical trials of antidepressants that we had not included in our meta-analysis, and there was also a meta-analysis of those other trials that had used some of the same methods we had used. It showed the same results that we had reported. The difference between drug and placebo in published trials of antidepressants was modest at best.[3] Still, the controversy continued.

In the midst of this dispute, I received a letter from Thomas J. Moore, a senior fellow in health policy at the George Washington University School of Public Health and Health Services. Noting the continuing controversy over our article, Moore proposed that I replicate our study with a different and more complete data set. He suggested that I use the US Freedom of Information Act to obtain the data that the drug companies had sent to the Food and Drug Administration (FDA) in the process of getting their drugs approved for marketing.

The FDA is the regulatory body that licenses medications in the US. The data submitted to it are the data that are submitted to regulatory agencies around the world – including the Medicines and Healthcare products Regulatory Agency (MHRA), which approves drugs for marketing in the UK, and the European Medicine Agency (EMEA), which licenses medications for the EU. So these were the data upon which the antidepressants that are on the market today were approved for doctors to prescribe. If there was anything wrong with those data, then arguably the drugs should not have been approved in the first place.

There are a number of advantages of analysing the FDA reports. One is that they include unpublished as well as published studies. Before approving medications, the FDA requires that the drug companies send them information on all of the trials that the company has conducted, regardless of whether or not those trials have been published. This is important because many clinical trials – especially those that have not been successful – are not published. A report by authorities at the Medical Products Agency (MPA) in Sweden suggests that as many as 40 per cent of clinical trials of antidepressants are not published.[4] In general, there is a tendency for successful studies to be published and for unsuccessful studies either not to be submitted for publication or to be rejected. This tendency is called 'publication bias', and it creates serious problems when one is reviewing the published literature. Because of publication bias, reviewers are likely to overestimate the effect of the drug they are reviewing. By gaining access to statistical summaries of the complete data set in possession of the FDA, my colleagues and I were able to avoid this publication bias.

A second advantage of using the FDA reports is that the agency carefully scrutinized the data that the drug companies had sent them. They examined the design of each of the studies and appraised the statistical procedures that were used to analyse the results. They asked the companies to provide more information and conduct additional data analyses where they deemed these to be needed. Most importantly, they excluded from consideration inadequate and poorly controlled trials. This enabled us to cope easily with one of the vexing problems of meta-analyses – that of assuring that all of the various studies included in the analysis were up to par. This part of our job had been done for us by a team of medical and statistical experts with the authority to gain information to which we had no access.

Finally, all of the trials in the FDA data set included the same measure of depression, a physician-rated scale called the Hamilton Rating Scale for Depression (HRSD). The Hamilton scale is completed by doctors based on interviews and observations of patients. The doctor rates the patient's mood, thoughts about suicide, sleep disturbances and other symptoms of depression. For example, one point is given if the patient feels that life is not worth living, and four points are scored if the person has made a serious suicide attempt. The result is a numerical score that can range from 0 to 51.

The virtues and shortcomings of the Hamilton scale can be debated, but it is a widely used scale with known clinical properties. The FDA uses it as its primary measure of drug effectiveness, the American Psychiatric Association (APA) has developed categories of severity of depression based on it, and NICE has used it to establish cut-offs for establishing clinical significance. Having Hamilton scores for the trials meant that we could interpret the meaning of the results in clinical as well as statistical terms. In other words, we could examine the effects of the drugs in terms of how meaningful they are in people's lives.

THE VANISHING DRUG EFFECT

Moore's idea of analysing the data that had been sent to the FDA seemed brilliant, and I proposed that we work on it together. So we began. Moore wrote to the FDA invoking the Freedom of Information Act and requested the medical and statistical reviews of every placebo-controlled clinical trial for the treatment of depression by what, at that time, were the six most widely used 'new-generation' antidepressant drugs: Prozac, Seroxat (Paxil in the US), Lustral (Zoloft), Effexor, Dutonin (Serzone) and Cipramil (Celexa). Except for Dutonin, which was withdrawn from the market after it was linked to cases of liver failure, these are still among the most widely prescribed antidepressants in the world.

Obtaining the FDA files turned out to be pretty easy, and with the data from their reports in hand, I asked two postgraduate students, Alan Scoboria and Sarah Nicholls, to work with me on the analysis. Together we calculated the degree to which people improved on each of the active drugs and how well they improved on placebos. Our first stumbling block was the discovery that there were missing data, even in the FDA medical and statistical reviews. We had data from all of the clinical trials for Prozac, Effexor and Dutonin, but not from some of the studies of Seroxat, Lustral and Cipramil. We knew of the existence of these clinical trials, because they were mentioned in the FDA documents. We also knew that they were 'adequate and well-controlled' trials, because they were described as such in the FDA reviews. Finally, we knew that they were negative trials – that is, they had not shown a significant difference between drug and placebo. This information was also included in the FDA files. What were missing were the actual numbers. For these particular clinical trials, we did not have the exact degree to which depression scores decreased after patients were given drug or placebo. Still, as Sapirstein and I had already shown and others have since confirmed, there is not much difference in the effectiveness of one antidepressant compared to another,[5] and we did have the complete data for Prozac, Effexor and Dutonin. Eventually, we were able to obtain the missing Seroxat data as well. As part of the settlement of a lawsuit against them by the State of New York, the manufacturer of Seroxat, GlaxoSmithKline, established a website on which they provide summaries of all their clinical trials. Using the information on this website, we later filled in the gaps in the FDA data set and redid our analysis. The results were the same either way. And even without the data from their worst trials, Lustral and Cipramil fared no better.

Analysing the data we had obtained from the FDA – data that

included unpublished as well as published studies – we found even less of a drug effect than in our analysis of the published literature.[6] Our analyses showed that 82 per cent of the response to medication had also been produced by a simple inert placebo. As conventionally interpreted, this means that less than 20 per cent of the response to antidepressant medication is a drug effect.

To put this into perspective, you might consider some calculations that my colleague Tom Moore has performed on some other data that he obtained from the FDA. These showed that about 50 per cent of the effects of a pain medication can also be produced by placebos, whereas the placebo effect in drugs used to treat blood-sugar levels is nil. In contrast, most of the improvement shown in drug-company trials of antidepressants was due to the placebo response. In fact, most of the clinical trials submitted by the drug companies failed to show any significant benefit of their drugs at all. More important, the average difference between improvement in the drug groups and improvement in the placebo groups was only 1.8 points on the Hamilton scale. The Hamilton is a 51-point scale, so a difference of less than two points is very small indeed. For example, one can get a six-point reduction in Hamilton scores merely by sleeping better, even if there is no other change in the person's depressive symptoms.

Having differences in Hamilton scores was particularly important because it meant that we could evaluate the clinical significance of the drug effect, as well as its statistical significance. When researchers report that a difference is significant, what they usually mean is that the difference is significant *statistically*. Statistical significance refers to whether an effect – the difference between a drug and a placebo, for example – is real, or whether it has just occurred by chance. It tells you how likely you are to get the same results if you do the same study over again. But it does not tell you how large or important the effect is. Whether a difference is statistically significant depends on a number of factors, including the number of people that were included in the study. The larger the study, the easier it is to find statistically significant differences. If the study is large enough, even very tiny differences will be statistically significant. Conversely, the smaller the study, the harder it is to find differences that are statistically significant. With very small studies, even relatively large effects might not be significant statistically. It is like a seesaw. When the size of the study goes up, the criterion for statistical significance goes down; and when the size of the study goes down, the criterion for statistical significance goes up.

To evaluate the importance of the difference of an effect, you have to look at the *clinical* significance of the findings. Unlike statistical significance, clinical significance refers to the size of the effect. It addresses whether it is likely to make a meaningful difference in anyone's life. An example might help clarify this. Imagine that a study has been conducted on 500,000 people and has found that smiling increases life expectancy. This seems very impressive, but on reading further you discover that it increases life expectancy by only ten seconds. With 500,000 subjects, the effect is likely to be statistically significant, but it is not clinically meaningful.

So how can we judge the clinical significance of the 1.8-point difference between improvement on antidepressants and improvement on placebos? One way is to look at the Hamilton scale and see how a difference of that size could be obtained. There are two common versions of the Hamilton scale: a 17-item version and a 21-item version. Fortunately, we do not have to be concerned about differences between these two versions, because only the first 17 items on the 21-item scale are generally scored. So as far as scores are concerned, the 17-item version and the 21-item version are identical.

The Hamilton scale is based on an interview with a doctor. The doctor completes the scale after the interview, indicating scores for such symptoms as depressed mood, feelings of guilt, thoughts of suicide, insomnia, and so forth. Total scores can range from 0 to 51. A two-point difference can be obtained by no longer waking during the night, *or* by no longer waking early in the morning, *or* by being less fidgety during the interview, *or* by eating better. Any one of these changes can make a two-point difference in a person's depression score, even if there are no changes at all in the person's depressed mood, feelings of guilt, suicidal thoughts, anxiety, agitation or any of the other symptoms of depression.

In my opinion – and in the opinion of just about everyone in the field to whom I have spoken – a two-point difference in depression scores on the Hamilton scale is not clinically meaningful. But we need not rely on my opinion. NICE has established a criterion for assessing the clinical significance of drug–placebo differences on the Hamilton depression scale.[7] According to NICE, the difference between drug and placebo has to be at least three points to be considered clinically significant. So the 1.8-point average difference in improvement that we found in the drug-company-sponsored trials of their products is quite far from being clinically significant.

DEPRESSION SEVERITY AND ANTIDEPRESSANT EFFICACY

As an invited speaker at various medical schools and hospitals, I have often been asked how severely depressed the patients in the drug-company clinical trials had been. Maybe antidepressants are no better than placebos for mildly depressed patients, it was suggested, but perhaps they work well for people who are severely depressed. In other words, the small average effect that we found might be misleading. It might hide a substantial effect for severely depressed patients that is masked by no effect at all for mildly depressed people. Indeed, the NICE guidelines concluded that there is some evidence of a clinically significant effect of the drugs in severely depressed patients, but not in those who are only mildly or moderately depressed.[8] NICE's conclusions were based on the published data, however, and my colleagues and I had the unpublished data as well. So we reanalysed the FDA data to see whether severity made a difference. To help with this project, I enlisted the aid of two experts on the theory and practice of meta-analysis, Professor Blair Johnson and his associate Dr Tania Huedo-Medina at the University of

Connecticut, as well as that of Dr Brett Deacon, a researcher at the University of Wyoming, who had identified the journal articles corresponding to those trials that had been published.

We examined the data in a number of ways. One was to use the classification system established by the APA to categorize levels of depression. The APA system, which was also adopted by NICE, divides scores on the Hamilton depression scale into the following five categories:

- No depression (0–7)
- Mild depression (8–13)
- Moderate depression (14–18)
- Severe depression (19–22)
- Very severe depression (23 and above).

In examining baseline depression scores (that is, measures of how depressed the patients were before the clinical trial began), the first thing we noticed was that all but one of the trials had been conducted with patients whose scores put them in the 'very severe' category of depression. The single exception was a clinical trial of Prozac conducted with moderately depressed patients. In other words, our findings of a clinically insignificant difference between drug and placebo was based primarily on data from those patients who are the most severely depressed according to the APA and NICE classification scheme.

There was no drug effect at all for the moderately depressed patients. They got considerably better when given antidepressants – in fact, mildly and moderately depressed people are the ones most likely to become completely free of depression when given treatment – but they showed just as much improvement when given placebos. Among the very severely depressed patients, there was a *statistically* significant difference between drug and placebo, but it was pretty much the same as the difference we had found when we had analysed the trials without regard for initial severity of depression. Removing the data for moderately depressed patients did not have much of an effect on the outcome of our analysis. The difference between drug and placebo was still less than two points on the Hamilton scale, well below NICE's three-point criterion for clinical significance. So the failure to find a clinically significant drug–placebo difference was not because the patients were only mildly depressed to begin with. The drug effect was small even for severely depressed patients.

Still, there was a relationship between severity and the antidepressant drug effect. Figure 2.1 shows that relationship. It indicates the amount of improvement that was shown at each level of depression severity. Now this is a rather complicated figure, so let me walk you through it. The triangles represent the drug response on each of the clinical trials; the circles indicate the placebo response. The size of the triangle or circle reflects the number of subjects in the trial. The larger the shape, the larger the trial. This is important because data from larger trials are more reliable than data from smaller trials. So when doing a meta-analysis, more weight is given to large trials than to small trials.

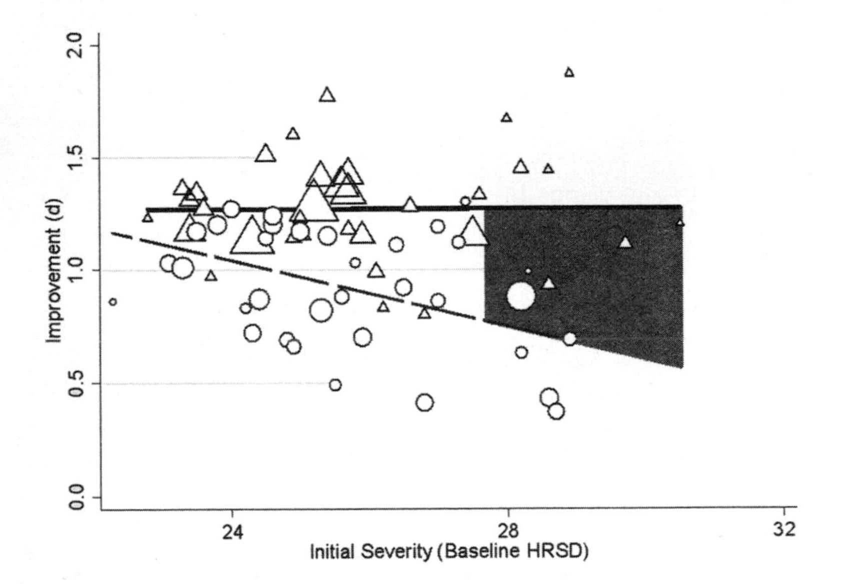

Figure 2.1. The response to drug and placebo at different levels of initial severity of depression.[9]

The most important things to look at in Figure 2.1 are the solid horizontal line, representing the average drug response, and the dashed diagonal line, representing the average placebo response. The difference between them is the drug effect. That difference gets greater and greater as baseline severity increases, until it finally reaches clinical significance (the shaded area) for the most extremely depressed patients – those with Hamilton scale scores of about 28 or more at the beginning of the study they were in. The average drug–placebo difference in this small group of relatively small studies was just over four points on the Hamilton scale. A four-point difference is clinically significant, according to NICE, but it is still rather small. Differences in sleep patterns, for example, can produce a six-point difference in depression scores, without any other differences in symptoms of depression. Still, this relationship seems to be reliable. The worse the depression, the greater the drug effect.

If you look at the figure again, you will see that there is something a bit strange about it. The response to the drug does not become greater as depression increases. Instead, the placebo response gets smaller, and that is what makes the drug–placebo difference larger. Now this seems very curious. Why is there less of a placebo response among extremely depressed patients, without much change in the drug response? I can think of two factors that might account for this. First, these patients tend to have been chronically depressed. They are much more likely to have been on antidepressant medication before, and they know what it feels like. Second, physicians are likely to prescribe higher doses to patients who are more severely depressed. As I show later in this chapter, the dose-response studies that have been done tell us that this does not make much difference in the effect of the antidepressant. Low doses of SSRIs are just as effective as higher doses. But unlike the therapeutic effects, the side effects of SSRIs *are* dose-dependent. The higher the dose of the medication, the more side effects you get. Putting these two factors together suggests that more extremely depressed patients are particularly likely to recognize whether they have been put on placebo or on the real drug. When they don't experience the side effects they are used to, even on high doses of the new medication, they may conclude that they have been placed in the placebo group, and this recognition may dampen the placebo effect. If this is the case, then even the relatively small but clinically significant drug effect seen in extremely depressed patients may be a placebo effect in disguise.

Now I have to admit that my speculation about the severity effect being due to breaking blind, and guessing correctly whether or not one has been given the real drug, is just conjecture. There may be other explanations. I have not been able to think of any, nor has anyone suggested to me another plausible explanation. Still, I consider my proposed explanation to be no more than a hypothesis, which might very well turn out to have been mistaken. But it might also be correct. And if it is, then there may not be a real drug effect, even amongst the most severely depressed patients.

A LITTLE GOES A LONG WAY

Prior to submitting our analysis of the published data to a journal, Sapirstein and I were concerned that we might have underestimated the drug effect by lumping together effective drugs with ineffective drugs – a concern that proved to be unfounded, as there turned out not to be any meaningful differences between one type of drug and another, even when looking at drugs that are not antidepressants. My colleagues and I had an analogous concern about the data we had received from the FDA. This time it was not differences in type of drug that concerned us – all of them were drugs that were supposed to inhibit the reuptake of the neurotransmitter serotonin – but rather differences in prescribed doses.

There are two ways in which clinical trials can be conducted. One method is to allow physicians to adjust the dose of the drug for each individual patient, just as they would in normal clinical practice. An inadequate response to treatment might lead the doctor to increase the dose. Concerns about side effects might lead to a lower dose. This is an excellent clinical-trial practice, in that it mimics what would happen when the drug is placed on the market. But it leaves an important question unanswered: what is an effective dose of the medication being tested?

To answer that question, a different type of clinical trial is used. In dose-response trials, patients are randomly assigned to receive low, moderate or high doses of the drug – or no drug at all in the placebo condition. Our concern was that patients given low doses of the antidepressant might not have responded because the dose was too low. By including these patients in our analysis, we might have underestimated the drug effect.

To check whether this might have biased our results, we compared the effect of treatment with the lowest dose of the drug to that of treatment with the highest dose. This led to the next of my many surprises. Putting the data from all of the dose-response trials together, we found that there was no difference between the effect of a high dose of antidepressants and the effect of a low dose. The average improvement on the Hamilton scale was 9.97 points on the highest dose of the drugs and 9.57 on the lowest dose.

Looking at the trials individually, we found 40 statistical comparisons between specific doses of the same drug. These yielded only one significant finding: low doses of Prozac were more effective than high doses. When you do a large number of statistical comparisons, you expect to get some spurious findings due to chance, and that is probably what the one test showing that a lower dose is better than a higher dose was – a chance finding. By and large, there is no relationship between how much of an antidepressant people take and how much they improve.

Some drugs produce effects at relatively small doses, following which it does not matter how much more you administer. A small dose of cyanide, for example, will leave you just as dead as a large dose. But most drug effects are dose-dependent. A small glass of wine at dinner has much less of an effect than four pints of lager afterwards. Even placebos have dose-related therapeutic effects. A Dutch researcher, Ton de Craen, and his colleagues found that ulcers healed at a significantly greater rate when patients were treated four times a day rather than twice a day, despite the fact that the treatment in both cases was a placebo.[10] But unlike alcohol or placebos, the therapeutic effects of antidepressants are not dose-dependent – at least not when the patients are unaware of whether they are getting a high dose or a low dose. Although higher doses of antidepressants can produce more side effects,[11] they do not produce greater reductions in depression. The difficulty of finding dose-related therapeutic effects of antidepressants is yet another reason for suspecting that those effects may be independent of their chemical action.

The equivalence of high and low doses of antidepressants is well known, yet doctors often increase the dose of the antidepressant when their patients do not improve. Why do they do this? The official Summary of Product Characteristics for Prozac provides a clue. It notes that 'in the fixed dose studies of patients with major depression there is a flat dose response curve, providing no suggestion of advantage in terms of efficacy for using higher than the recommended doses'. Nevertheless, despite the absence of evidence that higher doses produce better effects, the very same document advises physicians as follows:

> The recommended dose is 20mg daily. Dosage should be reviewed and adjusted if necessary, within 3 to 4 weeks of initiation of therapy and thereafter as judged clinically appropriate. Although there may be an increased potential for undesirable effects at higher doses, in some patients, with insufficient response to 20mg, the dose may be increased gradually up to a maximum of 60mg. Dosage adjustments should be made carefully on an individual patient basis, to maintain the patients at the lowest effective dose.

So when increasing the dose of antidepressants, doctors are merely following the manufacturer's advice, as reported in the Summary of Product Characteristics.

If the dose response curve is flat and higher doses produce an 'increased potential for undesirable effects', why does the Summary of Product Characteristics advise doctors to triple the dose if patients do not respond well enough to a lower dose? The key to understanding this contradiction is our old and trusted friend, clinical experience. The company notes that despite the negative data, 'it is clinical experience that uptitrating [increasing the dosage] might be beneficial for some patients'.

A study reported by Otto Benkert and his colleagues at the Department of Psychiatry at the University of Mainz shows how this works.[12] Depressed patients who failed to respond to antidepressant medication were given an increased dose of the drug, following which 72 per cent of them improved significantly by showing at least a 50 per cent reduction in symptoms of depression. The catch was that the dose had only been increased for half of the subjects. The others only thought the dose had been increased; in fact it had not. Yet the response rate was the same 72 per cent in both groups. So a patient whose dose of the drug is increased may indeed show more improvement, but this effect may be due to the patient's knowledge that the dose has been increased, rather than to the chemical effect of the medication. In other words, doctors are advised to increase the dose (and the likelihood of troubling side effects) as a means of strengthening the placebo effect.

SECRETS AND REVELATIONS

Our first published report of the FDA data was accompanied by nine expert commentaries, some of them by researchers who had conducted clinical trials of antidepressant medication. Although there were vast differences in interpretation, this time there were no doubts about the accuracy of our analysis. Some commentators argued that our analysis had actually *overestimated* the real effect of antidepressants. Others argued that the clinical trials sponsored by the drug industry are flawed and that they may underestimate the actual benefit of antidepressants. But all agreed that our description of the data was accurate. As one defender of antidepressants phrased it, '"the data are the data," and it is clear that antidepressants have relatively small, specific effects for the patients who participate in the RCTs [randomized clinical trials] conducted by the pharmaceutical industry.'[13]

After my experience with the previous meta-analysis, I was very pleasantly surprised by the consensus about our basic findings. I was even more surprised to learn that our findings did not come as news to those who were actively involved in antidepressant research. Indeed, one group of researchers wrote: 'Many have long been unimpressed by the magnitude of the differences observed between treatments and controls, what some of our colleagues refer to as the "dirty little secret" in the pharmaceutical literature.'[14]

So we had not discovered anything new at all. We had just uncovered a 'dirty little secret' that had been known all along. The companies that produce the drugs knew it, and so did the regulatory agencies that approve them for marketing. But most of the doctors who prescribe these medications did not know it, let alone their patients.

Pharmaceutical Companies Keeping Mum

How was this secret kept? How is it that even the doctors who prescribe antidepressants did not know how limited their effects were compared to dummy pills? Pharmaceutical companies have used a number of devices to make their products look better than they actually are. They have:

- Withheld negative studies from publication
- Published positive studies multiple times
- Published only some of the results from multi-site studies
- Published data that was different from what they submitted to the FDA.

The tendency to publish only the more successful trials has been most clearly documented by Hans Melander and his colleagues at the Medical Products Agency (MPA) in Sweden.[15] The MPA, which is responsible for approving new medications for marketing in Sweden, is the Swedish equivalent to the FDA in the US, the MHRA in the UK, and the EMEA in the EU. Melander and his colleagues searched the medical literature for the publications corresponding to the clinical trials that the drug companies had submitted to the MPA. They found that almost all of the successful clinical trials had been published, whereas most of the negative trials had not been published. Solvay Pharmaceuticals, for example, reported ten trials of the SSRI Faverin to the Swedish authorities. Only three of these studies showed a significant benefit for the active drug; the other seven did not. All three successful clinical trials were published as individual studies. Only one of the seven unsuccessful studies made it into print. In 1991 Faverin was approved for marketing in Sweden.

Earlier in this chapter, I mentioned this problem of publication bias – the tendency for successful studies to be published and unsuccessful studies not to be published. The failure to publish unsuccessful trials presents a problem in many research areas. When a study has produced non-significant results, it is less likely to be submitted for publication; and, if it is submitted, it is less likely to be favourably reviewed or accepted for publication. But although publication bias affects all areas of research to some extent, it is particularly acute when it comes to drug trials. This is because most of the clinical trials evaluating new medications are sponsored financially by the companies that produce and stand to profit from them. The companies own the data that come out of the trials they sponsor, and they can choose how to present them to the public – or to withhold them and not present them to the public at all. With widely prescribed medications, billions of dollars are at stake. In this case, it is not reviewers or journal editors who are impeding publication of negative findings. Rather it is the companies themselves that decide to withhold negative data from publication.

My contention that the drug companies sometimes withhold

publication of negative data intentionally is not merely an opinion or deduction. It is a documented fact.[16] During the 1990s, GlaxoSmithKline conducted three clinical trials on the efficacy of paroxatine, which is sold in the UK under the brand name Seroxat, in the treatment of major depression in children and adolescents. One study showed mixed results, a second showed no significant differences between drug and placebo, and the third trial suggested that the placebo might actually be more effective than Seroxat for children aged seven to eleven. Only one of these trials was ever published. The other two studies remained hidden, and the public might never have known about them, had a confidential internal company document not fallen into the hands of the *Canadian Medical Association Journal*.[17] According to the document, the company's 'target' was to 'effectively manage the dissemination of these data in order to minimize any potential negative commerical impact'. While acknowledging that the data were 'insufficiently robust', it nevertheless proposed the publication of the one study with mixed results. The company document noted that 'it would be commercially unacceptable to include a statement that efficacy had not been demonstrated, as this would undermine the profile of paroxetine'. So when the study was published in 2001, the article concluded that 'paroxetine is effective for major depression in adolescents'.[18]

In June 2004, Eliot Spitzer, who was then Attorney General and later became Governor of the State of New York, filed a lawsuit against GlaxoSmithKline, charging that the company had 'engaged in repeated and persistent fraud by concealing and failing to disclose to physicians information about Paxil [the brand name for Seroxat in the US]'.[19] The case was settled two months later, with the company agreeing to pay $2.5 million to the State and to establish an online clinical-trial register containing summaries of the results of all of the clinical trials they sponsored.[20] This is the website through which my colleagues and I were able to find the negative trial data that had been missing from the FDA files. Spitzer predicted that other manufacturers of antidepressants would soon follow suit, but by and large they did not.[21] Most studies showing negative results remain unpublished, and short of making official enquiries to government agencies, their data are unavailable to researchers, doctors and the public at large.

Picking Cherries and Slicing Salami

One would think that withholding negative data and publishing only the successful trials would be sufficient to maintain the 'dirty little secret'. But the pharmaceutical companies had other tricks up their sleeves. Whereas many of the negative trials were not published at all, some of the positive trials were published many times, a practice known as 'salami slicing', and this was often done in ways that would make it difficult for reviewers to know that the studies were based on the same data.[22] In some cases, the authors were different, and references to previous publication of the data were often missing. Sometimes there were minor differences in the data between one publication and another, as well as between the data as presented to regulatory agencies and the data as published. So a reviewer trying to summarize the data would be likely to count the positive data more than once.

Another trick was to publish only some of the data from a clinical trial, a manoeuvre that researchers call cherry-picking the data. Some clinical trials are conducted in more than one location. These are called multi-centre studies. Multi-centre studies make it easier to find sufficient patients to conduct the trial. They also make it easier to cherry-pick the data. For example, one multi-centre study of Prozac was presented to the FDA as showing a drug–placebo difference of three points on the Hamilton scale. When data from this clinical trial was published, the difference was reported as 15 points – a five-times increase in effectiveness. How was this magical augmentation of the benefits of Prozac accomplished? The full study was conducted on 245 patients. The published paper reported data from only 27 of these patients. In the published version, the data from the bulk of the patients were left out, making the drug seem much more effective than it really was.

Drug companies also publish 'pooled analyses' of the trials they have conducted. That is, they bundle together the results of different trials and analyse the drug–placebo difference across them. This is similar to the meta-analyses my colleagues and I have conducted, but with one important difference. Our meta-analyses, in common with most others reported in the scientific literature, are based on all of the studies that we were able to find. In contrast, the drug companies pick and choose which studies they wish to include in their pooled analyses. For example, GlaxoSmithKline submitted 15 clinical trials of Seroxat to Swedish regulators. In addition to being published individually – sometimes more than once – studies with positive results were also included in six different pooled analyses. Most of the studies with negative results were, of course, not included in the pooled analyses.

There is yet another way in which pooled analyses can hide negative data. Rather than not publishing the negative data at all, the companies can bundle them together with data from positive trials, so that the overall result is positive. By so doing, they can truthfully claim that they have published the data from a negative trial, while hiding the fact that those data showed no difference between drug and placebo. The article by the Swedish regulators showed that the data from about 20 per cent of clinical trials were not published at all. The data from another 20 per cent of the trials were bundled together with data from more successful trials, so that their negative results were hidden from view. Taken together, approximately 40 per cent of the data are kept out of sight.[23] Practices like this make antidepressant drugs seem much more effective than they actually are, and they also make it exceptionally difficult for reviewers to establish how effective the drugs really are.

Perhaps the best indication of the difficulties that are posed by selective publication is provided by the problems that NICE faced when drawing up its 2004 guidelines for the treatment of depression. Having read our meta-analysis of the FDA data, NICE contacted me in the hope of adding the unpublished data to their analysis of the published data. Although they wanted to include the unpublished as well as the published data, they did not want to include any of the data more than once, because that would have biased the results. So they asked me whether I could tell them which of the FDA trials had been published and what the publications were corresponding to each. There was no easy way to do this. NICE tried, but eventually gave up and reported in their guidelines that although they had planned to combine their data with ours, 'it was not possible to determine which of the FDA data had been subsequently published'.[24]

For doctors, researchers and policy setters to do their work properly, they need to have access to full and complete information. This can be done if pharmaceutical companies are required to do the following:

1 Register all clinical trials before they are started.
2 Make summary data publicly available for all completed trials and for trials that have been stopped prior to completion.
3 Describe the methods used in those trials at a level of detail that is at least comparable to what is found in scientific-journal articles.
4 Include references to prior publications and reports of data, so that they cannot be inadvertently counted twice by reviewers.
5 Make the raw data available, so that independent researchers and agencies can do their own analyses of them.

The first of these requirements has already been met. In 2004 the International Committee of Medical Journal Editors established a requirement that all clinical trials be registered publicly before they begin enrolling patients.[25] Any trial that has not been registered will not be considered for publication, at least not by the journals that have agreed to this policy. This is a strong enough threat to ensure compliance. It means that the existence of the trial will be known publicly – but not necessarily its outcome. Knowing the existence of a trial is helpful, but not enough.

The data coming out of the trial also need to be publicly available, along with details of the methods by which the data were collected and the publications that have resulted from it. Without this, researchers reviewing the clinical-trial literature will not be able to make accurate assessments of the efficacy of medications, and doctors prescribing drugs to their patients will not have sufficient information to make informed recommendations.

The importance of the fifth of my proposed requirements – that of having the actual data available, rather than just summaries – is highlighted by an experience that NICE had when preparing their 2004 guidelines for the treatment of depression. They would have liked to analyse the effects of age, gender and ethnicity on treatment outcome. This is important information in drawing up treatment guidelines, because the drugs might be more effective for some groups of patients than for others. According to Sir David Goldberg, who chaired the panel that wrote the 2004 guidelines for the treatment of depression, NICE requested this information from the drug companies, but the companies refused to release it. 'If they had, we could have run analyses,' he said. 'No chance!'

If there are subgroups of depressed patients who respond well to antidepressants, it would be very important to know who they are, so that antidepressant prescriptions could be targeted to them directly. However, this might not be something that the manufacturers of these medications would be eager to find out, because if there are some patients who respond better than average, then there must be others who respond worse. You can see this in our analysis of the data on severity of depression. The drug effect was a little better than average for the most extremely depressed patients, but it was non-existent for those who were moderately depressed. Knowing who responds to a drug and who does not is very important, not only so that responders can be given effective medication for their depression, but also so that the non-responders are not given drugs that have potentially serious side effects, but produce no therapeutic benefit for them at all.

Regulatory Agencies Keeping Mum

In our meta-analysis, more than half of the clinical trials submitted to the FDA showed no difference between drug and placebo. Most reviewers of the clinical-trials literature have not had access to unpublished studies and may not even know of their existence. But the FDA and other regulatory agencies around the world knew of these data. Nevertheless, their existence is not even mentioned in the product labels, information leaflets and official Summaries of Product Characteristics (SPC) of most antidepressants.

Why didn't the regulatory agencies inform the public that many clinical trials of antidepressants failed to show a significant benefit of the drug over placebo? In the case of the FDA, we know the reason. It was not just an oversight. Buried in the FDA files on the SSRI Cipramil, my colleagues and I found an internal FDA memo expressing the opinion that 'the provision of such information is of no practical value to either the patient or prescriber' and that it need not be included in the labelling for the drug. In the US, labelling information is published in the *Physicians' Desk Reference*, a compendium of FDA-approved information to which physicians often turn in deciding what drugs to prescribe. So the decision to exclude information about failed clinical trials meant that most doctors prescribing the medication would not know of them.

The author of the FDA memo revealing the decision to hide the existence of the clinical trials that had failed to find a difference between drug and placebo was Dr Paul Leber, Director of the FDA Division of Neuropharmacological Drug Products. Here is what he wrote about the labelling of Cipramil, which is referred to by its generic name, citalopram, in the document:

> One aspect of the labeling deserves special mention. The Clinical Efficacy Trials subsection within the Clinical Pharmacology section not only describes the clinical trials providing evidence of citalopram's antidepressant effects, but makes mention of adequate and well controlled clinical studies that fail to do so. I

> am mindful, based on prior discussions of the issue, that the Office Director is inclined toward the view that the provision of such information is of no practical value to either the patient or prescriber. I disagree. I believe it is useful for the prescriber, patient, and 3rd party payer to know, without having to gain access to official FDA review documents, that citalopram's antidepressant effects were not detected in every controlled clinical trial intended to demonstrate those effects. I am aware that clinical studies often fail to document the efficacy of effective drugs, but I doubt the public, or even the majority of the medical community, are aware of this fact. I am persuaded that they not only have a right to know, but should know. Moreover, I believe that labeling that selectively describes positive studies and excludes mention of negative ones can be viewed as being potentially 'false and misleading.'[26]

Perhaps it is not without reason that in Italy a patient information leaflet is sometimes called a '*bugiardino*' – literally, 'little liar'.

Leber's laudable argument that mention of negative trials be included in the labelling information had relatively little effect. Although a brief mention of these trials was added to the citalopram label, the FDA continued to approve antidepressant labelling that did not disclose the existence of negative data. In fact, it went even further. It urged the drug companies to keep the studies hidden. According to an article in the *Washington Post*:

> The Food and Drug Administration has repeatedly urged antidepressant manufacturers not to disclose to physicians and the public that some clinical trials of the medications in children found the drugs were no better than sugar pills, according to documents and testimony released at a congressional hearing yesterday. Regulators suppressed the negative information on the grounds that it might scare families and physicians away from the drugs, according to testimony by drug company executives. For at least three medications, they said, the FDA blocked the companies' plans to reveal the negative studies in drug labels.[27]

How can we explain such strange behaviour on the part of regulatory agencies? Perhaps part of the answer lies in the way they are funded. Once upon a time, the FDA was funded solely by the government. But that changed in 1992, when Bush senior signed a bill into law allowing the FDA to charge the drug companies fees to evaluate their new products, so that they could be approved more quickly. One of the stipulations of that bill was that none of the funds could be used by the FDA to monitor the safety of the medications it approved. That was relaxed to some extent when the law was reauthorized, first under Bill Clinton in 1997 and again under George W. Bush in 2007, but it is still the case that only a small percentage of the fees can be used for safety monitoring.

In April 2007, when the law approving drug-company funding of the FDA was up for renewal, it was heavily criticised in a spirited article in the *New England Journal of Medicine*.[28] In that article, Jerry Avorn, a Professor of Medicine at Harvard University, described some of the effects that the law had on the FDA regulatory process. The law had been conceived in response to complaints by AIDS activists about how long it took for new drugs to be approved, and one of its provisions was the imposition of strict deadlines for decisions. In the FDA's efforts to meet those deadlines, its Office of Drug Safety was downsized, as resources were shifted from safety monitoring to drug approval. According to Avorn, 'One FDA scientist who was often criticized for being too concerned about drug-risk data was told by his supervisor to remember that the agency's client was the pharmaceutical industry. "That's odd," the FDA scientist replied. "I thought our clients were the people of the United States."'[29]

Lest you think the financial entanglement between the drug industry and those who regulate it is only an American problem, let me assure you that it is not. Drug-company funding accounts for 40 per cent of the FDA budget, but it provides more than 70 per cent of the income for the EMEA, and ever since Margaret Thatcher took the Department of Health out of the business of regulating the drug companies, all of the funding for the MHRA has come from the pharmaceutical industry.[30] It may not be a case of the fox guarding the hen house, but it does seem to resemble asking the thieves to feed the guard dog. If a conflict of interest is to be avoided, it might be better to fund regulatory agencies from general tax funds, even if the companies are then charged fees by the government to compensate for the expense.

WHY WERE THE DRUGS APPROVED? VOODOO SCIENCE

Although doctors and their patients did not know about the unpublished negative trials or how small the drug effect was, the regulatory agencies did. So how is it that these drugs were approved for marketing in the first place? This is an obvious question, and I am not alone in raising it. Officials from the drug regulatory agencies of the European Union, France, the Netherlands and Sweden raised the same question in a 'regulatory apologia' that they published after our 2008 meta-analysis came out. 'Against this background,' they wrote, 'one can ask why the new-generation antidepressant medicinal products were ever approved.'[31]

To answer this question, the regulators conducted their own meta-analysis of some of the data in their files. Their results were very similar to ours. They agreed that the average observed difference in improvement between drug and placebo is only about two points on the Hamilton scale, and their data also showed that most of the drug response could be explained as a placebo effect. Nevertheless, they argued that they had shown that the drugs were better than placebos, not only statistically, but clinically as well.

How were the European regulators able to pull off the trick of turning a two-point difference on the Hamilton depression scale into a clinically significant benefit? They did so by using a different criterion for improvement than the one we had used in our analysis of the clinical-trial data. We had analysed the average degree of improvement in symptoms that patients given antidepressants and placebos had experienced. This is a common measure of the drug effect, and it was used by NICE in establishing a criterion for clinical effectiveness. In their apologia, however, the European regulators examined response rates instead of average improvement. A 'response rate' is the percentage of people whose symptoms decreased by some specified amount. In antidepressant drug trials, a 50 per cent reduction in symptoms is most often used as the criterion for separating 'responders' from 'non-responders,' and this is the criterion that the European regulators used.

Using this common definition of response, the European regulators reported that 49 per cent of the people in the drug groups had gotten better, compared to only 33 per cent of patients given placebos. The difference between these two percentages is 16 per cent. The regulators argued that these patients had benefited from having been given the real drug instead of placebos; and that, they said, is clinically significant.

There are two ways of looking at these data. On one hand, they indicate that antidepressants only help 16 per cent of the patients to whom they are prescribed. The rest of those who get better would also have gotten better on a placebo. On the other hand, given the popularity of antidepressants, 16 per cent represents a lot of people, and one could well argue that a medication that can help so many people deserves to be marketed.

Still, this does not tell the whole story. The conclusion that 16 per cent of depressed people benefit only from the real drug is actually an illusion based on a numerical sleight of hand – although I suspect that the regulators were not aware of this when writing their apologia. My colleague Joanna Moncrieff and I have shown how the response-rate illusion works.[32] People whose symptoms have diminished by 49 per cent or less are classified as 'non-responders', and those whose symptoms improved by 50 per cent or more are classified as 'responders'. The illusion lies in the implication that the 'non-responders' have not gotten better at all. In fact, many of them have experienced substantial clinical improvement, so much so that a very small boost – as little as one additional point of the 51-point Hamilton depression scale – can push them over the 50 per cent criterion, turning them into 'responders'. These are the 16 per cent of depressed people who get classified as 'responders' when given an antidepressant and as 'non-responders' when given a placebo.

In other words, even the small percentage of people who 'respond' only to the real antidepressant do not get much chemical benefit from the medication. Most of their improvement can be explained as a placebo effect. On average, the drug adds two additional points of improvement on the Hamilton scale, beyond what these patients would have obtained on placebo. This is enough to push them over the arbitrary, but widely used 50 per cent criterion that separates clinical 'responders' from 'non-responders'. The boost might derive from a small specific effect that drugs have on depression, but as I noted in Chapter 1, it could also come from the experience of side effects from the active drug, which lead clinical-trial patients to conclude that they have been assigned to the drug group rather than the placebo group, thereby producing an enhanced placebo effect. But in either case the effect is small and it does not meet conventional criteria for clinical significance.

So why were the drugs ever approved? The real answer to this question lies in the criteria that are used for antidepressant drug approval. The efficacy criterion used by drug regulators requires two 'adequate and well-controlled' clinical trials showing that a drug is better than a placebo. But there are some catches. The first catch is that there is no limit to the number of studies that can be run in order to find the two showing a statistically significant effect. Negative trials just don't count. The second catch is that the size of the drug–placebo difference – its clinical significance – is not considered, although the published 'apologia' suggests that this might change.

The FDA approval of citalopram (Cipramil) provides a convenient example of how this works. Seven placebo-controlled efficacy trials were conducted. Two showed small but significant differences between drug and placebo. Another two trials failed to show significant differences, but were deemed too small to count. Three other trials that were deemed adequate and well controlled also failed to show significant drug–placebo differences. For each of them, the leader of the FDA new drug application team stated that 'the reasons for the negative outcome for this study are unknown', and for two of them he added that 'there was a substantial placebo response, making it difficult to distinguish drug from placebo'. In his summary of these three negative trials, the team leader wrote, 'I feel there were sufficient reasons to speculate about the negative outcomes and, therefore, not count these studies against citalopram.'[33] Agreeing with this assessment, the Division Director concluded that 'there is clear evidence from more than one adequate and well controlled clinical investigation that citalopram exerts an antidepressant effect'.[34] This, in my opinion, is voodoo science. The drug companies can conduct as many trials as they want until they find two showing significant effects. The negative trials simply don't count.

Even with the five negative trials discounted, the size of the drug–placebo difference for citalopram was small. It averaged only two points on the Hamilton scale, well below the three-point difference that NICE uses as its criterion for clinical significance. The FDA reviewers recognized that the difference was small. The team leader wrote that 'while it is difficult to judge the clinical significance of this difference, similar findings for other SSRIs and other recently approved antidepressants have been considered sufficient to support the approvals of those other products'. So citalopram was approved as well.

THE ASSAY SASHAY

When reading the FDA memos on citalopram, I was struck by a curious phrase. The FDA team leader for the new drug application wrote that the three adequate and well-controlled negative trials were 'not easily interpretable since there were no active control arms'.[35] An 'active control arm' is a group of subjects who are given an older established drug, against which the effects of the new drug might be evaluated. So there are 'two-arm' trials, in which a drug is compared to a placebo, and 'three-arm' trials, in which a new drug is compared to both a placebo and an older drug.

Why would you include an older established drug in a clinical trial? You might suppose that this would be to see if the new drug is better. That is what I had assumed until I read the FDA memo, but the memo made me wonder. Why should the absence of another drug in the study make it harder to interpret a failure to find a difference between the new drug and a placebo?

It turns out that there is another reason for including a second medication in a clinical trial. You may remember that there is not much difference in effectiveness between one drug and another in the treatment of depression (see Chapter 1). So trying to show that the new drug works better is not likely to pay off, and given how expensive clinical trials are and how difficult it can be to recruit subjects for them, a drug company would not want to include an additional 'arm' unless it paid off. So why include it?

Here is how it works. Let's say the new drug successfully outperforms placebos in the study. That's fine; it means that the drug works. In this case, it really does not matter that it is not more effective than the old drug, and it does not matter whether the old drug worked better than placebos. But what if the drug does not work significantly better than placebos? That is what happens in about half of the clinical trials in which antidepressants and placebos are compared. In that case, you can look at whether the old drug did better than placebos. If it did not, then you conclude that the study lacked 'assay sensitivity'. An assay is an analysis or assessment. So if a trial lacks assay sensitivity, it means that it is not sufficiently sensitive to analyse the effectiveness of the drug, and that therefore the study should not be counted as evidence against the new drug. The logic is as follows: we know the old drug works, because it has already been approved. So if this clinical trial doesn't show it to be effective, it must be that the trial is not 'sensitive' enough to detect differences. No matter that the old drug did not work in many of the trials that led to its approval, and there is no need to explain or even speculate as to why the study was not sensitive enough. Assay sensitivity is just a way to stack the deck in favour of the new drug.

The assay sashay is like betting on coin tosses with the following rules. We toss two coins. If the first one comes up heads, I win, and the second coin is irrelevant. If the first one comes up tails, we have to decide whether the toss counts. To do that, we look at the second coin. If it also comes up tails, the toss does not count and we call it a draw. With these rules, I will win 50 per cent of the time and it will be a draw 25 per cent of the time. You win only if both the first coin comes up heads and the second comes up tails, which will only happen 25 per cent of the time. So the odds are heavily stacked in my favour. If you doubt this, please get in touch. I will be happy to play you for real money. Using these standards to judge the effectiveness of a medication is voodoo science to the nth degree.

* * *

Our analyses of the FDA data showed relatively little difference between the effects of antidepressants and the effects of placebos. Indeed, the effects were so small that they did not qualify as clinically significant. The drug companies knew how small the effects of their medications were compared to placebos, and so did the FDA and other regulatory agencies. The companies found various ways to make the data seem more favourable to their products, and the FDA helped them to keep their negative data secret. In fact, in some instances, the FDA urged the companies to keep negative data hidden, even when the companies wanted to reveal them. My colleagues and I hadn't really discovered anything new. We had merely revealed the 'dirty little secret'. How were our revelations received? In the next chapter I review the responses – favourable and unfavourable – that were aroused by the publication of our analyses, and I respond to the criticism that some defenders of antidepressant medication have levelled.

3

Countering the Critics

At first, I was very surprised to find that the response to antidepressant drugs was so small when compared to placebos. The media also found it new and revelatory – not only once, but three times. Ten years ago the meta-analysis of the published clinical-trial data that Guy Sapirstein and I had done was reported in newspapers, magazines, and television documentaries around the world, as well as in the news section of *Science*, one of the world's top scientific journals, spanning all branches of science. The first analysis that my colleagues and I had done on the FDA data set was also covered widely in the media. Nevertheless, I was unprepared for the reaction to our most recent and most complete analysis. I woke up on the morning of 26 February 2008 to find that it was front-page news in *The Times*, *The Guardian*, *The Independent* and the *Daily Telegraph*. It was reported on the BBC, ITV, Sky News and Channels 4 and 5. It made its way into newspapers and television and radio news programmes in the US, Spain, Portugal, Germany, Italy, South Africa, Australia, Canada, China and many other countries. It was also reported and debated in a number of leading medical and scientific journals. Overnight, I seemed to have been transformed from a mild-mannered university professor into a media superstar – or super-villain, depending on whom you asked.

In addition to attracting media attention, the publication of the 2008 meta-analysis also had practical effects. On 23 May 2008, a scant three months after its publication, Onmedica.com published a survey of 490 doctors in the UK, in which it asked them what effect our analysis would have on their prescribing practice. Almost half (44 per cent) said that they would change their prescribing habits and consider alternative treatments rather than SSRIs for their depressed patients.[1]

It is not often that researchers find their work leading to such widespread changes of behaviour. Still, the 44 per cent figure reveals a split opinion. Most physicians did not intend to alter their prescribing practices. Our analysis has provoked a vociferous and continuing debate on the effectiveness of antidepressants and the circumstances under which they should be prescribed. In this chapter I consider and respond to the various criticisms that have been levelled at our data-based conclusions about the efficacy of antidepressants.

'ANTIDEPRESSANTS WORK IN CLINICAL PRACTICE'

Many doctors and patients have reacted to our meta-analysis with simple disbelief. David Nutt, head of the Psychopharmacology Unit at the University of Bristol, said, 'Antidepressants work in clinical practice – everybody knows they work.' Another critic wrote:

> Dozens of clinical trials plus decades of clinical practice plus millions of content patients can't be that wrong. Whatever the bias in whatever the study, common sense clearly says: the sum of the parts attesting antidepressants' efficacy blatantly outnumbers the evidence showing the opposite. The use of these antidepressants is now deeply rooted and well-established in medical society worldwide, it's safe, it works, and there's no shadow of doubt about it.[2]

In a way, these critics are right. Clinical experience does show that prescribing antidepressant drugs works – and so did our meta-analyses. Patients given antidepressants in the clinical trials we analysed showed substantial, clinically meaningful improvement. But so did those given placebos, and the difference between the drug response and the placebo response was not great. The question is not whether antidepressants work, but *why* they work. Is it because the chemical in the pill specifically targets depression, or is it because of the placebo effect?

Physicians do not sysematically prescribe placebos to their patients. Hence they have no way of comparing the effects of the drugs they prescribe to placebos. When they prescribe a treatment and it works, their natural tendency is to attribute the cure to the treatment. But there are thousands of treatments that have worked in clinical practice throughout history. Powdered stone worked. So did lizard's blood and crocodile dung, and pig's teeth and dolphin's genitalia and frog's sperm. Patients have been given just about every ingestible – though often indigestible – substance imaginable. They have been 'purged, puked, poisoned, punctured, cut, cupped, blistered, bled, leached, heated, frozen, sweated, and shocked',[3] and if these treatments did not kill them, they may have made them better.

Because of the power of the placebo effect, almost anything that is believed in seems to work for some types of medical problems. That is why the late Arthur K. Shapiro described the history of medicine as largely the history of the placebo effect.[4] It is also why clinical experience alone cannot tell us whether a particular physical substance is an effective treatment. Placebo-controlled trials are required to demonstrate drug efficacy before drugs are approved for marketing.

The problem is that many doctors are extremely reluctant to drop treatments that seem to work in clinical practice, even when clinical trials show that these treatments are really placebos. This problem is not confined to antidepressants. As we shall see in Chapter 5, it is a problem that has slowed medical progress in other areas as well.

CLINICAL PRACTICE VERSUS CLINICAL TRIALS: THE STAR*D TRIAL

There are a number of ways in which clinical practice is different from the clinical trials we analysed, and these differences are often cited in efforts to dismiss our findings. One difference is that the patients in the clinical trials knew they might be given a placebo. As I described in Chapter 1, knowing that the pill one is taking might be a placebo decreases its antidepressant effect.[5] However, this knowledge also reduces the effectiveness of a placebo. Placebos are more effective when people are misled into believing that what they are getting is definitely a powerful active treatment than when they are told that they might be getting a placebo.[6] If knowing that one might be getting a placebo decreases the response to both the placebo and the drug, then the net effect of this knowledge on the drug–placebo difference should be zero.

Another difference between clinical trials and clinical practice is that each of the patients in the clinical trials we analysed was given only one kind of treatment. When a patient seen in clinical practice fails to respond to a particular antidepressant, psychiatrists often prescribe a different one. Sometimes the second antidepressant works. When it doesn't, a third might be prescribed and then a fourth and a fifth, until one is found that works. The implicit logic behind this practice is that different patients suffer from different chemical imbalances. Some people may be depressed because they have a shortage of the neurotransmitter serotonin in the brain; SSRIs, which are supposed to selectively target serotonin, should work fine for them. Others might be lacking in norepinephrine as well as serotonin and would best be served by an SNRI, a drug that enhances the availability of both types of neurotransmitter. Still others might have perfectly adequate serotonin levels, but might be lacking in norepinephrine and dopamine; they would need a medication such as bupropion that targets these two neurotransmitters. In other words, one has to find the right drug for the right patient. We might call this the tailoring hypothesis, as the task of the physician is to tailor the treatment to the particular chemical imbalance that is causing each individual patient's depression.

The fact that patients sometimes improve when they are switched from one antidepressant to another is often interpreted as evidence for the tailoring hypothesis. It also leads to the claim that antidepressants are effective, despite the clinical-trial evidence showing rather small effects. The particular antidepressant given to patients in any one clinical trial may have been the right drug for some of them, but the wrong drug for others. Maybe that is why the drug effects do not seem very large when the average effect on all of the patients is calculated. The clinically observed phenomenon of patients recovering after being switched from one antidepressant to another suggests to many doctors that finding the right drug for the right person might be the key to antidepressant efficacy.

The clinical experience that patients sometimes improve when a different medication is prescribed has received confirmation from a very unusual and widely heralded clinical trial. The study is called the Sequenced Treatment Alternatives to Relieve Depression – or STAR*D – trial.[7] It was designed to be more representative of what happens in 'real world' clinical practice than are typical clinical trials, and also to show the effectiveness of antidepressants in the best of circumstances. A broader range of patients than are included in normal clinical trials was accepted into the STAR*D study, there was no placebo control group, and – most importantly to our present discussion – patients who did not get better on the first drug were given a different treatment. Those who were still depressed after being given a second medication were switched to a third, and those not responding to the third were given a fourth.

As I noted in Chapter 2, there are different ways to measure the outcome of a clinical trial. One way is to examine how much better the patients have gotten – that is, the average degree to which their symptoms have been reduced. Another common method is to analyse how many of the patients have gotten better. Most often, a 50 per cent reduction in symptoms is used as a criterion of what is called a 'clinical response'. One problem with this method is that most of the severely depressed patients who show a clinical response are still depressed at the end of the clinical trial – despite the fact that their depressive symptoms have been cut in half.[8] This is part of the response-rate illusion that I talked about in Chapter 2. It makes a treatment look more effective than it really is. The researchers who designed the STAR*D trial used a much more stringent criterion. They examined the number of patients who were in remission, meaning that they were no longer depressed at the end of the trial.

Using this very strict criterion of remission, the STAR*D researchers reported that 37 per cent of the patients in the trial recovered from depression on the first medication they were given. Another 19 per cent of the full group of patients recovered on the second medication, 6 per cent on the third and 5 per cent on the fourth. Altogether, 67 per cent of the patients recovered. However, the remission of symptoms turned out to be only temporary for many – approximately half of the patients who recovered relapsed within a year.

This is a rather bleak picture of the effects of antidepressant treatment. In the best of circumstances – which is what the trial was designed to evaluate – only one out of three depressed patients showed a lasting recovery from depression, and since there was no evaluation of what the recovery rate might have been with placebo treatment, there is no way of knowing whether their recovery was actually due to the medication they had been given.

Still, the study did seem to show that switching from one antidepressant to another might make a difference. But does it? To understand the real significance of the STAR*D trial, it is helpful to consider a much older study.[9] In 1957, a team of researchers at the University of Oklahoma School of Medicine gave ipecac – a drug that is used to induce nausea and vomiting – to a group of volunteer subjects. After verifying that ipecac did indeed elicit nausea and vomiting in these subjects, the researchers then gave them a treatment to prevent nausea and vomiting, followed by ipecac again. The question was: would the treatment inhibit the nausea and vomiting that ipecac induces? As in the STAR*D trial, they repeated this procedure with different medications, in this case switching medications regardless of whether the previous one had worked. They did this seven times, and on each occasion they measured the success of the treatment at preventing nausea and vomiting.

The Oklahoma study showed the same pattern of results as the STAR*D trial. Some treatments seemed to be effective for some patients and other treatments seemed effective for others. More than half of the subjects responded successfully to the first treatment; 17 per cent did not respond to the first treatment, but did respond to the second. The third treatment was successful for another 20 per cent who had not responded to prior treatments, and by the time the sixth treatment was tried, 100 per cent of the subjects had successfully responded to at least one of them.

Like the STAR*D trial, this study seemed to show that different people respond to different medications and that the key might be finding the right treatment for the right person – but there was a catch. None of the medications were real treatments for nausea or vomiting. Instead, they were all placebos.

With the Oklahoma study in mind, we can reconsider the meaning of the STAR*D data – and the meaning of what happens in clinical practice when doctors switch medications. The results of the STAR*D trial might have had nothing to do with switching antidepressants. Instead, they might have been due to the placebo effect, which, as the Oklahoma study had shown, can kick in at any time. They could also have been due to other factors as well. Relief from depression may have occurred because of changes in the patients' lives, or simply because levels of depression tend to fluctuate over time. Furthermore, these various possibilities are not mutually exclusive. There may have been one reason for the improvement that one patient experienced and another reason for another patient's improvement. The point is that the patients might have gotten better even if they had been switched to a placebo, as was done in the Oklahoma trial, or even if they had been given nothing at all. Similarly, when patients in clinical practice get better after having been switched to a second antidepressant, it may have nothing to do with the change in medication. Instead, it could be the placebo effect kicking in – perhaps because the patient knows that a different treatment is being used – or it could be due to natural fluctuations in the course of their depression.

The idea of tailoring treatment to the patient implies that the benefit of switching drugs derives from different chemical imbalances that might be causing the depression. But the data from the STAR*D trial contradicts this, even without taking into account the results of the Oklahoma study. The first drug that was given to all patients in the STAR*D study was an SSRI. SSRI stands for 'selective serotonin reuptake inhibitor', which means that the drug is supposed to inhibit the reuptake of serotonin, but not of other neurotransmitters. If this did not work, patients were given one of three different antidepressants. Some of the non-responsive patients were switched to an SNRI, a drug that blocks the reuptake of the neurotransmitter norepinephrine, in addition to blocking serotonin. Others were given bupropion, which does not affect serotonin at all, but instead inhibits the reuptake of norepinephrine and dopamine. A third group of patients was simply switched to another SSRI, the same type of drug to which they had not responded in the first place.

Switching non-responsive patients from an SSRI to an SNRI led 25 per cent of them to get better. Change from an SSRI to bupropion produced virtually the same remission rate (26 per cent). But what of the patients who were not switched to a different class of antidepressant, but instead were simply given another SSRI? Twenty-seven per cent of these patients also got better – a remission rate that is virtually identical to that produced by changing to a different type of medication. In other words, the rate of improvement did not depend on the kind of drug to which the patient had been switched. Simply changing from one SSRI to another was as effective as changing to a completely different type of antidepressant. Once again we have the strange 'coincidence' of virtually identical effects produced by chemically different drugs. This indicates that it is not the specific chemical action of the drug that alleviates the person's depression. Instead, it may simply be the idea of changing treatment.

CLINICAL TRIALS ARE FLAWED

The most common criticism of our meta-analysis is the claim that the clinical trials we analysed were flawed, and that better results would have been found if the studies had been designed better. The trials were too short to show the real effect of antidepressants, the critics said. The people recruited to participate in them were not depressed enough, or they were too depressed. In any case, they were not representative of the patients who are generally seen in clinical practice.

Taken as a whole, these seem like rather strange criticisms for proponents of antidepressants to raise. These were the trials that were the basis upon which the drugs were approved. If there was anything seriously wrong with these studies, then arguably the drugs should not have been approved in the first place. Furthermore, the studies were sponsored by the drug companies. One would expect them to have been designed to maximize the benefit shown by the products in which the companies had invested so much money.

In fact, studies funded by drug companies usually show positive effects of their products and worse results for the products of their competitors, whereas studies that have been independently sponsored show results that are midway between these two extremes. A team of researchers at the Beth Israel Medical Center in New York have examined the outcome of clinical trials as a function of who had sponsored them. They found that approximately 75 per cent of drug-company studies showed favourable results for their own drugs, but only 25 per cent of them showed favourable results for the product of a competing company. In studies that are not sponsored by a drug company, the success rate is approximately 50 per cent.[10] So it seems rather unlikely that the industry-sponsored studies we evaluated would underestimate the drug effect. Still, we should look at each of the alleged flaws of these clinical trials and see whether they might have led to an underestimation of the efficacy of antidepressants.

The Trials Were Too Short

The clinical trials from which efficacy is gauged are relatively short. Most of the trials we analysed were only six weeks long, although some of them lasted eight weeks and a few were only four weeks long. Perhaps this is not long enough to show the real drug effect.

It is widely believed that the drug effects of antidepressants take two to three weeks to become evident, and that any improvement seen before then is likely to be a placebo effect. Still, four to eight weeks should give plenty of time for a drug effect to be seen. Furthermore, the belief that the therapeutic effects of antidepressants are delayed is based on clinical experience, and a recent meta-analysis of the clinical-trial data contradicts it.[11] The analysis, conducted by researchers at the University of Oxford, Yale University and the University of Birmingham, showed that the largest decrease in depressive symptoms occurred by the end of the first week of treatment. Although improvement continued for at least six weeks – the typical length of a clinical trial – the rate of improvement was less each week, and during the last couple of weeks of the trials the difference between drug and placebo did not seem to increase at all. The authors of the study commented that their evidence seemed to exclude 'the possibility that treatment response from antidepressant drugs is subject to a period of delay'. So increasing the length of clinical trials beyond the usual four to eight weeks is not likely to increase the drug effect.

Defenders of antidepressants cite 'relapse prevention' or

'discontinuation' trials to support their contention that the real magnitude of the drug effect requires longer trials to become evident. Relapse-prevention trials are designed to assess what happens when patients are taken off medication. They work like this: patients who have responded reasonably well to the active medication are either kept on the drug or switched to a placebo. Then relapse rates are compared. Relapse-prevention trials generally show that switching patients to a placebo leads them to get worse, compared to those who stay on the active drug.

Now there are a number of problems with relapse-prevention studies. One is the fact that many people who are taken off antidepressants experience withdrawal symptoms, which in severe cases can last for months. Some of these withdrawal symptoms – sadness, suicidal thoughts, crying spells, trouble concentrating, irritability, anxiety, agitation and insomnia, for example – are also symptoms of depression.[12] These withdrawal symptoms could lead both patients and researchers to think that the patient has relapsed.

In addition to being mistaken for a relapse, antidepressant withdrawal symptoms might induce a real relapse. Imagine that you are a depressed patient who has been helped by an antidepressant. As part of a relapse-prevention study, the drug is withdrawn and you are given a placebo. Shortly thereafter you begin to feel sad, depressed, anxious and agitated. These were all symptoms that you experienced when you were depressed. In addition, you now begin to experience some new symptoms that you have never felt before. You feel nauseous, dizzy, your vision blurs and your muscles twitch. Not knowing that all of these are symptoms of antidepressant withdrawal, you may think you have relapsed and become even worse than you were before beginning medication. Misinterpreting withdrawal symptoms as an indication of relapse could initiate a vicious cycle leading to a genuine relapse.

A second problem with relapse-prevention trials is related to research suggesting that the use of antidepressants might make people more vulnerable to relapse. Patients who are being treated with antidepressants show a specific vulnerability to relapse that is not shown by recovered patients who have been treated without drugs.[13] So the relapses suffered in relapse-prevention trials may be due to a biological vulnerability that has been induced by the medication in the first place. In the next chapter, where I review the evidence behind the myth that depression is a disease caused by a chemical imbalance in the brain, I also discuss in more detail the studies indicating that antidepressants might make people more vulnerable to relapse.

Finally, one of the concerns that I raised about clinical drug trials earlier in this book is that differences in side effects might enable patients to figure out whether they have been put in the drug group or the placebo group. This problem is especially salient in relapse-prevention trials. Imagine that you have agreed to be a subject in one of these studies. You have already been in the shorter efficacy trial, and you have been told that you were in the active drug condition. You have experienced the drug effect, and you have also experienced some of the side effects produced by the active medication. Now you are told that you may be continued on the active drug or you may be switched to a placebo. Isn't it likely that you would be able to detect at least some difference if you were switched to a placebo, even if only a difference in side effects? So much for double-blind! Rather than the discontinuation of medication, it may be the patients' knowledge that medication has been discontinued that causes them to relapse in these studies.

There is a reason for most efficacy trials being relatively short. As time goes on, patients tend to drop out of them, either because the drug is not working well enough or because of side effects. When too many patients have dropped out of a trial, the study is considered to have been compromised and its validity is called into question. So short-term trials are the norm.[14]

Nevertheless, some long-term efficacy trials have been conducted. These 'continuation' studies are different from relapse-prevention or discontinuation trials in some very important ways. Instead of just looking at patients who have responded to the active drug, continuation trials also look at patients who responded to a placebo. People who have gotten better on the drug are kept on the drug, and those who have gotten better on the placebo are kept on the placebo. Because no one is switched from drug to placebo or from placebo to drug, it is less likely that the patients in continuation trials will figure out which they have been given.

In 2002, the prestigious *Journal of the American Medical Association* published a six-month clinical trial comparing the SSRI Seroxat to a placebo to St John's wort, a herbal remedy that has been widely used to treat depression, particularly in Germany, where it is a registered substance for the treatment of mild to moderate depression.[15] The first part of the study was a short-term efficacy trial that lasted eight weeks. Those who got better were asked to stay on whichever treatment they had been given for another 18 weeks, bringing the total length of treatment to six months. This should certainly be long enough to show a difference between drug and placebo – if there is one.

At the end of the first eight weeks there was no significant difference between any of the groups. Patients in all three groups had improved substantially, regardless of whether they had been given the SSRI, the herbal remedy or the placebo. At the end of six months there was still no difference between groups. Those who had improved on the active drug maintained their improvement, but so did those who had improved on the placebo. In fact, only one patient in the entire study relapsed, and that was a subject who had been given the herbal remedy.

In case you think that six months may not be long enough, note that similar results were shown in a year-long industry-sponsored continuation trial comparing two different antidepressants (Seroxat and imipramine) to placebo. In that study, patients who had responded to either of the antidepressants or to the placebo during the initial six-week trial were kept on their treatment for an additional year. Patients in all three groups maintained their improvement. In fact, at the end of the one-year period, those who had been treated by placebos were the least depressed, although the differences between the groups were small and could easily have been due to chance.[16]

These continuation trials tell a very different story from that told by relapse-prevention trials. They show that there is little difference between antidepressant and placebo even when the clinical trial is extended over a longer period of time. Across the eight continuation trials that have been published, 79 per cent of patients on placebo and 93 per cent of patients on active medication remained well throughout the treatment period. In these long-term studies, placebo treatment was 95 per cent as effective as drug treatment. The authors of a meta-analysis of these trials concluded that 'the widely held – and probably erroneous – belief that the placebo response in depression is short-lived appears to be based largely on intuition and perhaps wishful thinking'.[17]

When drafting their guidelines for the treatment of depression, NICE also reached the conclusion that there is little difference between drug and placebo in long-term studies. In addition to analysing short-term trials, they conducted a separate analysis of published studies that had lasted longer. They concluded that 'in trials lasting eight weeks or longer, there is evidence suggesting . . . a statistically significant difference favouring SSRIs over placebo on reducing depression symptoms . . . but the size of this difference is unlikely to be of clinical significance'.[18]

The two meta-analyses of long-term efficacy trials were limited to data that had been previously published, as are most meta-analyses. Nevertheless, the differences between drug and placebo were clinically insignificant. We can only wonder whether there are also some unpublished long-term trials that the pharmaceutical industry has sponsored, and if so, what the results were. There is one thing of which we can be fairly certain. Unpublished trials, where they exist, do not show any better results than the published trials. We can be certain of this because drug companies publish their successful studies, often many times over. It is the unsuccessful trials that remain unpublished.

That fact that what gets published are the trials with positive results was most convincingly shown by a group of researchers at the Oregon Health and Science University, who followed up on our initial analysis of the FDA data by comparing the conclusions reached by the FDA with those reported by the drug companies in journal articles. Of 38 drug-company clinical trials that the FDA viewed as having positive results, all but one was published. In the same documents, the FDA described 36 other trials as having negative or questionable results. Most of these negative trials were not published at all, and of the few that were published, most were described in the journal articles as showing positive results – despite the fact that the FDA had concluded that they had not.[19]

The Subjects Were Not Depressed Enough

One of the most surprising criticisms of our most recent analysis of the FDA data is that the patients in these trials were not depressed enough to show a strong drug effect. Antidepressants are not effective for mildly depressed patients, the critics noted, but they are for those who are severely depressed. This criticism is surprising because the whole point of our article was to analyse the extent to which the effects of antidepressants might depend on how severely depressed the patients were to begin with. What we found was that all but one of the trials involved patients who were classified as very severely depressed – the most severe category of depression used by the American Psychiatric Association and by NICE. So what is the basis for the concern that the patients in the trials were not depressed enough to show a strong drug effect?

The most common answer to this question is that the researchers fudge the data – that the doctors who assess the patients' levels of depression rate them as being more depressed than they actually are, so that they will qualify to be enrolled in the trial. It can be difficult to find enough patients for a clinical trial, and the doctors may be paid for each patient they enrol. So they can be under considerable pressure to qualify as many patients as they can.

Fudging the data is a very serious charge. If it is true, then the real response to drug treatment is even less than the clinical trials indicate (unless, of course, the researchers also rate the patients as being more depressed at the end of the trial – but why would they do that?). How seriously should we take the charge that doctors falsify the clinical-trial data? One way of assessing the likelihood that the researchers have fudged the data might be to consider who it is that is levelling this charge. One of these sources is Dr Janet Woodcock, Director of the FDA Center for Drug Evaluation and Research, who was quoted in an interview with a reporter from *USA Today* as saying that 'patients may be rated more ill than they really are at the outset because doctors are so eager to get them into drug trials.'[20]

Dr Woodcock went on to say that 'we [the FDA] make sure these drugs work before we put them on the market', but her charge that the doctors in clinical trials intentionally distort the data undermines this claim. If the doctors are under pressure to falsify the data at the beginning of a trial, would they not be under similar pressure to falsify them at the end of the trial, so as to make the patients look less ill after taking the antidepressant? Of course, the trials are double-blind, so how would the evaluators know which patients should be assigned lower scores? As we have seen, most doctors are able to figure out which patients are in the drug group and which are in the placebo group,[21] so the task of fudging the outcome data would not be that difficult. These trials are the basis for drug approval. If the data have been distorted in any way, then something is seriously wrong with the drug approval process.

Fortunately, there is a methodological feature of these clinical trials that makes the charge of falsifying the initial data somewhat less problematic than it might otherwise be – as long as those are the only data that were fudged. All of the trials we analysed had what is called a placebo 'run-in' or 'wash-out' phase. The way this works is as follows. After people are assessed for inclusion in the trial, they are all given a placebo for a week or two. After this run-in period, the patients are reassessed, and anyone who has improved is excluded from the trial. The baseline severity scores that we used in our analyses were those taken after the placebo run-in period; they were not the depression ratings that doctors had made at the very beginning of the patients' treatment. So these were patients who were rated as still being very severely depressed after two weeks of placebo treatment.

Of course, if the doctors distorted the baseline scores upon which we had based our classification of severity, they may also have fudged their second ratings of the patients' levels of depression. They may have fudged the initial data to enrol the patients in the first place, and then a second time to keep them in the trial after the placebo run-in period. But this makes the charge of fudging the data even more troubling. If the data have been intentionally distorted at least twice – first at enrolment and then again at baseline – then how can we trust the outcome assessments that were made at the end of the trial?

The placebo run-in period might itself distort the clinical-trial data, but not in the direction that the critics of our meta-analyses contend. By getting rid of the patients who show a placebo response to the medication that is being investigated, the run-in should make the drugs look more effective than they actually are. This is a potential flaw in clinical-trial methodology that biases the trials in favour of the drug and against the placebo. It is also an ethically questionable practice.[22] Patients are not told that they will definitely be put on a placebo for a while, nor are they told this at the debriefing at the end of the study. So the placebo run-in is a violation of the requirement for informed consent.

Sandra Lee and her colleagues at St Boniface General Hospital in Winnipeg, Canada, analysed the effect of including a placebo run-in period and excluding patients who get better from the trial. Although placebo run-ins tended to produce larger drug effects, the difference was not statistically significant. Part of the problem was that there were relatively few studies that did not have a run-in phase. This makes it very difficult for a difference to reach statistical significance. The authors concluded that 'from a scientific standpoint, there is no reason to use the placebo run-in phase to eliminate placebo responders because it is costly in terms of time and effort'. But they also predicted that the practice of using placebo run-in periods will continue, because they do in fact seem to produce larger drug effects.[23]

The Subjects Were Too Depressed

While some critics have complained that the patients in the clinical trials we assessed were not depressed enough, others have argued that they were too depressed. An editorial in *Nature Reviews Drug Discovery*, for example, complained that 'all but one trial analysed involved groups with mean initial depression scores in the "very severe" range, limiting the strength of extrapolations'.[24] Their suspicion seems to be that antidepressants are more effective for severely depressed patients, less so for either the mildly depressed or very severely depressed groups, then becoming more effective again for the most extremely depressed.

This is a rather labyrinthine possibility. It suggests that, like the serpentine body of the Loch Ness monster, which rises and falls above and below the water in drawings that you see of it, the effects of antidepressants might rise and fall below and above the threshold of clinical significance, depending on how severely depressed the person was to begin with.

In Chapter 2 I described an analysis by European drug regulators of the antidepressant data that the drug companies had submitted to them.[25] Their analysis shows that the concern raised by the editors of *Nature Reviews Drug Discovery* was ill-founded. The regulatory agency data included trials of severely depressed patients, as well as trials with moderately depressed and very severely depressed patients, and since it included unpublished as well as published trials, it was perfectly suited to answer the question of whether severely depressed patients might be more responsive than those who are either more or less depressed. As I described in the previous chapter, their study showed a relatively small but statistically significant benefit for antidepressant drugs compared to placebo, which is exactly what we had concluded in our meta-analyses. More important to the issue at hand, the European regulators found no evidence at all that responses to antidepressant drugs were linked to severity of depression. Patients who were severely depressed were no more likely to respond to antidepressants (or to placebos for that matter) than those who were either moderately depressed or very severely depressed.

The Patients Were Not Representative

Many patients are excluded from clinical trials. Critics of our meta-analysis have suggested that antidepressants might work better for these patients than they do for those who are studied in clinical trials. Let us see how plausible this concern is.

One of the main reasons for excluding patients from clinical trials is to make it easier to find differences between the drug and the placebo.[26] There are two ways in which excluding some patients from the trials can help accomplish this aim. One is to eliminate those who are most likely to respond to a placebo. To accomplish this goal, patients are excluded from clinical trials if they have only been depressed for a short time, if they are only mildly depressed or if they respond to placebo treatment during the placebo run-in phase.

The second way in which excluding patients from clinical trials can magnify drug–placebo differences is by getting rid of patients who are not likely to respond well enough to the active drug. This is the reason for excluding patients who have been depressed for a very long period of time, who have not responded to previous treatments, who abuse alcohol or other drugs or who, besides being depressed, also suffer from an anxiety or personality disorder or from various medical disorders. Patients with these characteristics do not seem to respond as well as others to drug treatment. The exclusion criteria used in clinical trials make it difficult to know exactly how well antidepressants work in the broader population of depressed patients in clinical practice, but if there is a bias, it favours the drugs rather than the placebo, because the purpose of excluding these patients is to increase drug–placebo differences so as to make the drug effect easier to detect.

Most clinical trials, including the ones my colleagues and I analysed, are conducted on volunteers, many of whom are recruited for the trial by advertisements. Perhaps these depressed people are not as responsive to antidepressants as the patients seen in clinical practice. The STAR*D trial that I described earlier was designed specifically to evaluate the effect of antidepressants on the kinds of patients who are typically seen in clinical practice. None of the patients in this trial were recruited by advertising. Instead, they were all patients who sought treatment for depression in family practice or psychiatric out-patient treatment facilities. Also, the usual exclusion criteria were relaxed, so that a broader range of patients was evaluated. The trial did exclude patients who had already tried antidepressants but had not responded to them, although this exclusion should result in better response rates, not worse ones.

If the critics of our analyses of the drug-company trials were right, the STAR*D trial ought to have shown a larger drug response than that which is typically reported in clinical trials. In fact, it did not.[27] Instead, it showed remission and response rates very similar to those reported in placebo-controlled clinical trials like the ones my colleagues and I had analysed. This despite the fact that the STAR*D trial did not include a placebo control group, a feature that has been shown to increase responsiveness to antidepressant treatment.[28] If anything, the clinical trials my colleagues and I analysed showed better results than trials with a more representative group of patients would have shown.

The Burden of Proof

Critics of our analyses who claim that the trials are flawed implicitly assume that it is my task to prove that antidepressant drugs do not work. But where does the burden of proof lie? Earlier in this chapter I listed a few of the substances that have been used medicinally over the centuries. Others include putrid meat, fly specks, human sweat, worms, spiders, furs and feathers. These treatments seemed to work in the past at least well enough for doctors and their patients to have had confidence in them, and we cannot prove that they do not work, because they have not been tested in clinical trials. So perhaps we should go back to using them until appropriate trials prove their ineffectiveness. After all, as the critic of our meta-analysis that I quoted at the beginning of this chapter put it, 'clinical practice plus millions of content patients can't be that wrong'.[29]

The point is that the practice of medicine should be based on empirical evidence, not on its absence. I do not have to prove that antidepressants do not work. Instead, it is the job of the drug companies to prove that they do work. If the trials were flawed, then clinically significant differences between antidepressant and placebo have not been established for most patients. If the trials were not flawed, the data indicate that 'clinically significant differences between antidepressant and placebo have not been established for most patients' (quoted from the previous sentence). Either way, the objective of proving the effectiveness of antidepressant medication has not been met.

Furthermore, it is not enough to show that antidepressants are statistically better than placebos. For drugs to be marketed and for patients to be exposed to their side effects and other risks, the benefit over placebos needs to be shown to be clinically significant. In the files I obtained from the FDA, agency officials acknowledged the failure to show a clinically significant benefit for the drugs they have approved, saying instead that they demonstrate 'proof in principle' of the effectiveness of the drugs. What proof in principle means is simply that the drugs are statistically superior to placebos, even if the difference is vanishingly small. In approving Cipramil (Celexa in the US), the Director of the FDA Division of Neuropharmacological Drug Products summarized the situation as follows:

> The size of [the] effect, and more importantly, the clinical value of that effect, is not something that can be validly measured, at least not in the kind of experiments conducted. Accordingly,

> substantial evidence in the present case, as it has in all other evaluations of antidepressant effectiveness, speaks to proof in principle of a product's effectiveness.

And the Team Leader for Psychiatric Drug Products commented, 'While it is difficult to judge the clinical significance of this difference, similar findings for other SSRIs and other recently approved antidepressants have been considered sufficient to support the approvals of those other products.' In other words, the 'clinical value' of an antidepressant drug is just not part of the FDA's criteria for approving it.

But do the clinical-trial data submitted to the FDA even establish proof of principle? Recall that the rather small differences found between drug and placebo in the trials submitted to the FDA could have been due to the breaking of blind on the basis of perceived side effects. It may simply be evidence of an enhanced placebo effect, rather than a true drug effect. As I noted in Chapter 1, once side effects are taken into account, the difference between SSRI and placebo is not even statistically significant.[30]

Although the FDA does not consider the clinical value of an antidepressant when approving it, NICE did take clinical significance into account when drafting their clinical guidelines. More recently, European regulators, in their 'apologia' following the publication of our most recent meta-analysis, acknowledged that clinical relevance should be a consideration in drug approval, and they tried – unsuccessfully in my opinion – to show retrospectively that the data demonstrate clinical as well as statistical significance. This is a welcome change, and it is one that I hope will come to be adopted as official policy.

SUBSEQUENT TRIALS SHOW DIFFERENT RESULTS

By and large, the drug companies reacted reasonably well to the various analyses that we conducted on their data. Two of the companies actually hired me for brief consultations. They were not at all surprised by our findings – they knew it all along, and they wondered what all the fuss was about. In fact, the reason they hired me was that the placebo effect was so strong. They were finding it difficult to demonstrate drug effects and were hoping to find a way to identify in advance those people who were likely to respond to placebo treatment. If they could accomplish this, they could exclude the 'placebo responders' from clinical trials, and with these people excluded it might be easier to show a drug effect.

GlaxoSmithKline (GSK), the manufacturer of Seroxat, seems to be the only drug company that has commented publicly on our meta-analysis.[31] We had limited ourselves to analysing the data submitted to the FDA, which included 16 trials of Seroxat. They have conducted more than 170 trials, they said; so the 16 trials we had analysed were just a small proportion of the studies they had done.

As I described in Chapter 2, following a lawsuit in which GSK was accused of hiding some of the negative clinical-trial data, the company was required to maintain a website that reports the results of all of its studies of Seroxat. I have examined the studies that GSK has put on its website. Most of them do not have placebo control groups. They are irrelevant to the argument of whether SSRIs are much better than a placebo for the treatment of depression. But the GSK website also reveals some placebo-controlled 'post-marketing' studies – that is, studies that were conducted after the FDA had approved Paxil (as Seroxat is called in the US). Because our analysis was limited to the FDA data set, we had not included these later studies. Do they tell a different story?

Coincidentally, researchers at a World Health Organization (WHO) centre, the University of Verona in Italy and the Nagoya City University in Japan had already analysed all of the placebo-controlled antidepressant trials on GSK's website, and published their results at just about the same time that we published our analysis of the FDA data. They found 40 placebo-controlled studies of Seroxat for the treatment of major depression, including the 16 that had been sent to the FDA. The results of their analysis of these 40 studies were virtually identical to the results of our analysis of the studies that had been sent to the FDA. We had found that placebos were 82 per cent as effective as antidepressants in treating major depression. In the later study, which included the post-marketing studies on GSK's website, the placebo was 83 per cent as effective as the real drug.[32]

OIL AND WATER OR GUNS AND KNIVES?

There is yet another possibility. The general assumption is that the effect of a drug adds to the placebo effect, so that the total improvement that patients experience is the drug effect in addition to the placebo effect. This assumption is implicit in the design of placebo-controlled clinical trials, in which the drug effect is assessed as the difference between the response to the drug and the response to the placebo. Anne Harrington, an historian of science at Harvard University and the London School of Economics, calls it the oil-and-water hypothesis.

However, drug effects and placebo effects may not be additive like oil and water.[33] They could be independent, so that the response would be the same even if there were no placebo effect at all. Instead of being like oil and water, drugs and placebos may be like guns and knives. Shooting someone will leave the victim just as dead as shooting him and stabbing him, and the fact that stabbing a person leaves her just as dead as shooting her doesn't mean that shooting is ineffective. Similarly, it is possible that the effects of antidepressants are real and that they are large, despite the small size of the difference between drug and placebo. Maybe depressed patients would get better when given antidepressants even if they were given the drug without knowing it. In other words, the response to antidepressant medication could be a true drug effect that is masked by the placebo effect in clinical trials.

A number of studies with healthy volunteers have tested whether various drug and placebo effects are additive. These studies use an experimental method called the 'balanced placebo design'.[34] Figure 3.1 shows how this type of study is done. Half of the subjects in the study (those in groups A and B in the figure) are told that they have been assigned to the drug group. The others (groups C and D) are told that they are in the placebo group. Sometimes this information is true; sometimes it is not. The subjects in group A have been given a drug and know that they have been given the drug. Those in group B think they have been given the real drug, but have actually received a placebo. In group C subjects have been given a drug, but think they have just taken a placebo. Group D is a control group that shows what happens when people are given nothing at all, not even a placebo.

		Told they are getting:	
		Drug	Placebo
Actually get:	Drug	A	C
	Placebo	B	D

Figure 3.1 The balanced placebo experimental design

The balanced placebo design makes it possible to assess whether or not drug and placebo effects are additive like oil and water, or whether the placebo merely masks effects that are really being produced by the drug. If drug and placebo effects are additive, subjects who are knowingly getting the real drug (those in group A) ought to improve more than those who are either getting a placebo (group B) or getting the drug without knowing it (group C). On the other hand, if there is a real drug effect that is being obscured by the placebo effect – that is, if the two effects are not additive – then people given the drug without knowing it (group C) ought to do better than those who do not get drug or placebo (group D), even if drug and placebo effects (groups B and C) are equivalent.

Studies using methods like the balanced placebo design indicate that some drug effects are additive and some are not. For example, drug effects and placebo effects add together to produce a stronger combined effect when assessing the effects of caffeine on alertness or the effects of morphine on pain.[35] But not all drug and placebo effects add together in this way. Sometimes, some rather strange interactions are found. For example, the tenseness or jitteriness that some people feel when they drink too much coffee only happens when people consume caffeine and know that they are consuming caffeine (group A in the figure). Caffeinated coffee does not make people jittery when they think the coffee is decaffeinated (group C), and placebo caffeine does not make them jittery either (group B).

There is some indirect evidence suggesting that the antidepressant drug effect – if there is one – and the placebo effect are additive. As I described in Chapter 1, patients in clinical trials in which there is no placebo condition improve significantly more than patients in placebo-controlled trials. In the placebo-controlled trials, patients are told that they might be given a placebo, and this knowledge diminishes the effect of taking the drug.[36] Still, the pure drug effect of antidepressants has not been assessed in a balanced placebo study, and it is possible that a test of this sort would reveal a larger effect than that shown in typical clinical trials.

Given that possibility, you might think that the drug companies would be eager to try a study of this sort. In fact, they are not. I have been campaigning for a direct test of the additivity hypothesis for years,[37] but the drug companies do not seem to be inclined to sponsor a trial that could accomplish this goal, or if they have, the results have not been published. Although the results of such a test might vindicate antidepressants and show that they work independently of the placebo effect, they could also confirm that antidepressants are little more than active placebos. Why take the chance?

* * *

In the process of writing this book and responding to the various concerns raised by critics of our meta-analysis, I have come across data of which I had not previously been aware – some of which had not been published at the time my colleagues and I wrote up our meta-analysis for publication. These data indicate that we were overcautious in our interpretation of the data we had received from the FDA.

Until now, I have argued that the therapeutic effects of antidepressants are small, that clinically meaningful benefits *may* be limited to a small subset of patients, and that the effects of the drugs *may* not be due to their specific chemical composition – that instead of being active therapeutic agents, antidepressants may instead be active placebos. The process of addressing the objections of my critics has steadily driven me to a set of much more far-reaching conclusions. I have always had my doubts about the commonly held view that depression is a brain disease – a chemical imbalance that is reversed by antidepressant medication. Now, considering all of the data together, I have come to believe that the chemical-imbalance theory is completely implausible.

In the next chapter I examine the data behind the chemical-imbalance theory. Others have argued that these data provide only weak support for this conventional view.[38] I go a step further. I do not think the data are weak at all. They are in fact rather strong. But rather than supporting the chemical-imbalance theory of depression, they contradict it. It now seems beyond question that the traditional account of depression as a chemical imbalance in the brain is simply wrong.

4

The Myth of the Chemical Imbalance

Depression, we are told over and over again, is a brain disease, a chemical imbalance that can be adjusted by antidepressant medication. In an informational brochure issued to inform the public about depression, the US National Institute for Mental Health tells people that 'depressive illnesses are disorders of the brain' and adds that 'important neurotransmitters – chemicals that brain cells use to communicate – appear to be out of balance'. This view is so widespread that it was even proffered by the editors of *PLoS* [Public Library of Science] *Medicine* in their summary that accompanied our article. 'Depression,' they wrote, 'is a serious medical illness caused by imbalances in the brain chemicals that regulate mood', and they went on to say that antidepressants are supposed to work by correcting these imbalances.

The editors wrote their comment on chemical imbalances as if it were an established fact, and this is also how it is presented by drug companies. Actually it is not. Instead, even its proponents have to admit that it is a controversial hypothesis that has not yet been proven.[1] Not only is the chemical-imbalance hypothesis unproven, but I will argue that it is about as close as a theory gets in science to being disproven by the evidence.

HOW THE BRAIN WORKS

To understand the chemical-imbalance theory, it will be helpful to first review some basic aspects of how the brain functions. The human brain contains about 100 billion nerve cells called neurons. Each neuron is like an electrical wire with many branches. When a neuron fires, electrical impulses travel along its length from one end to the other. When an impulse reaches the end of a branch, it may stimulate the next neuron, influencing whether or not it fires.

Neurons do not actually touch each other. Rather, there are fluid-filled gaps, called 'synapses', between the end of one neuron and the beginning of another. The brain's electrical impulses are not strong enough to span these gaps. So how can a neuron's electrical impulse influence the firing of a neighbouring nerve cell? It does so by means of chemicals called 'neurotransmitters', which are manufactured by neurons and convey information across the gaps between them (that is, the synapses). Serotonin is one of the neurotransmitters through which one neuron influences the firing of another. Others include norepinephrine and dopamine. There are many other kinds of neurotransmitters, but these three – and especially serotonin – have been hypothesized to be involved in depression.

After neurotransmitter molecules have influenced the firing of a receiving neuron (more technically called a postsynaptic neuron), some of them are destroyed by enzymes in the synaptic cleft (the synapse), some are reabsorbed by the sending presynaptic neuron in a process that is called 'reuptake', and the rest remain in the space between the two neurons. The chemical-imbalance hypothesis is that there is not enough serotonin, norepinephrine and/or dopamine in the synapses of the brain. This is more specifically termed the monoamine theory of depression, because both serotonin and norepinephrine belong to the class of neurotransmitters called monoamines.

INVENTION OF THE CHEMICAL-IMBALANCE THEORY

The 1950s gave rise to the Korean War, the Cuban revolution, the Hungarian revolution, the hydrogen bomb, beatniks – and antidepressants. Two different types of antidepressants were developed during this decade, and in both cases the discovery of apparent antidepressant effects was serendipitous. The story of how antidepressants were discovered – or perhaps 'invented' might be a better word – and how they led to the development of the chemical-imbalance theory is rather convoluted.[2] But it is worth examining, as there are important lessons to be learned from it. From the beginning, the chemical-imbalance theory was based on weak and contradictory evidence, and data contradicting it were simply ignored. This is a pattern that was to be repeated. A half-century of research has produced data indicating that the chemical-imbalance theory must be wrong. Yet it remains the most popular explanation of depression, and most of the data contradicting it continues to be ignored.

The first antidepressant was a drug called iproniazid that had been produced in 1951 from leftover German rocket fuel by the pharmaceutical company, Hoffmann-La Roche, and was being used for the treatment of tuberculosis. As is true of most medications, clinical trials of iproniazid revealed various side effects, but not all of these effects were negative. Some patients reported an increased sense of vitality and well-being. At first, this was merely considered a side effect and was ignored, but it was not long before clinicians in France and the United States began trying iproniazid as a treatment for depression.

In 1957, Nathan Kline, Harry Loomer and John Saunders, at the Rockland State Hospital in Orangeburg, New York, reported the first influential assessment of iproniazid as a 'psychic energizer' on non-tubercular psychiatric patients, some of whom were suffering from depression. According to their report, about two-thirds of patients showed a 'measurable response' to the drug. This is about the same response rate that is reported for clinical trials of antidepressants today, and as we have seen, most if not all of that response can be attributed to the placebo effect. But the study conducted by Kline and his colleagues did not include a placebo control group – placebo-controlled clinical trials had not yet become fashionable – and the antidepressant effect was assumed to be a biological response to the drug. In less than one year, more than 400,000 depressed patients had been treated with iproniazid, and the first antidepressant had been born.[3]

One year after Kline and his colleagues reported the effect of iproniazid on psychiatric patients, a Swiss psychiatrist named Roland Kuhn published an article in the *American Journal of Psychiatry* on the antidepressant effects of the tricyclic drug imipramine. Like iproniazid, the discovery of imipramine as an antidepressant was accidental. Kuhn was studying the effect of imipramine on psychosis, not depression, but three of his patients who had been diagnosed with psychotic depression showed marked improvement, and Kuhn went on to try imipramine on other depressed patients.[4] He reported that a high percentage of his patients recovered completely, usually within two to three days of being given the drug.[5] This is quite remarkable, given the subsequent widespread belief that it takes weeks for antidepressants to take effect.

It is important to note that claims for the effectiveness of iproniazid and imipramine were not based on placebo-controlled clinical trials. Instead, they were based on clinical impressions.[6] In 'discovering' the antidepressant effects of imipramine, Kuhn did not even use precise measurement, rating scales or statistics. His claim was that precise measurement led to stagnation rather than progress in medicine, and he preferred to rely on his extensive medical experience and 'artistic imagination' instead.[7]

Despite the weakness of the data, the idea that iproniazid and imipramine were effective antidepressants came to be widely accepted. This is not really surprising, in the context of the times. In the 1950s and 1960s, the power of the placebo effect was just beginning to be recognized, and placebo-controlled clinical trials were rare. New treatments were often accepted on the basis of clinical experience and the testimony of experts in the field.

Iproniazid and imipramine seemed to work as antidepressants, but how did they achieve their effects? It would be another decade before the chemical-imbalance theory was launched. In 1965, Joseph Schildkraut at the National Institute of Mental Health in Washington, DC, published a groundbreaking paper in which he argued that depression was caused by a deficiency of the neurotransmitter norepinephrine in the gaps between neurons in the brain.[8] Two years later Alec Coppen, a physician at West Park Hospital in Surrey, published another version of the chemical-imbalance theory. His version differed from Schildkraut's in that it put most of the blame on a different neurotransmitter, emphasizing serotonin rather than norepinephrine as the neurotransmitter that was lacking.[9]

What was the scientific basis for these chemical-imbalance theories? As I noted above, norepinephrine and serotonin are now known to be neurotransmitters – chemicals that transmit nerve impulses from one neuron to another. But in the 1950s knowledge of neurotransmission was sketchy at best. The presence of norepinephrine in the nervous system was not demonstrated until 1954, and evidence that dopamine functions as a neurotransmitter was not reported until 1958. As late as 1960 the idea that neurotransmission is largely chemical in nature, though advocated by a group of largely British scientists, was not yet widely accepted.[10]

Against this backdrop, researchers reported evidence that iproniazid, the antitubercular drug that was to become the first antidepressant, might increase norepinephrine and serotonin levels in the brain. How did it have this effect? Recall that some of the neurotransmitter molecules released by a neuron are destroyed by enzymes in the synaptic cleft between the sending presynaptic neuron and the receiving postsynaptic neuron. When the neurotransmitter is a monoamine – like norepinephrine and serotonin – this process is called monoamine oxidase (MAO). As early as 1952 researchers at the Northwestern University Medical School in Chicago reported that iproniazid inhibited the oxidation of monoamines.[11] This meant that iproniazid was a monoamine oxidase inhibitor – an MAOI, as this type of antidepressant is commonly called.

Here then is the logic behind the first version of the chemical-imbalance theory. Iproniazid is a monamine oxidase inhibitor – it inhibits the oxidation of norepinephrine and serotonin in the synapses, thereby leaving more of these neurotransmitters available in the brain. When depressed people take iproniazid, they get better. Therefore insufficient norepinephrine and/or serotonin causes depression.[12]

There was a problem with this first version of the biochemical theory of depression. Iproniazid was not the only drug that had been reported to be effective as an antidepressant. Imipramine, the drug that had been tested by the Swiss psychiatrist Roland Kuhn, seemed to have similar effects. But imipramine is not an MAOI; it does not inhibit the destruction of neurotransmitters in the synapse. So if antidepressants worked by inhibiting monoamine oxidase, why was imipramine effective? How could its apparent effectiveness be reconciled with the chemical-imbalance theory?

The answer is that there are two ways in which neurotransmitter levels might be increased. One is to inhibit their destruction after they have been released into the synaptic gap. That is how MAOIs are supposed to work. Recall, however, that after a neurotransmitter is released, some of its molecules are reabsorbed by the presynaptic neuron that released them in a process that is called 'reuptake'. Blocking this reuptake process should also increase the level of neurotransmitters in the brain. In 1961, Julius Axelrod, who later received the Nobel Prize in Medicine for his work on the release and reuptake of neurotransmitters, reported that imipramine, as well as a few other drugs, inhibited the reuptake of norepinephrine in cats. Two years later he reported that these drugs also inhibited the reuptake of serotonin.[13]

Axelrod's discovery provided an answer to the question of why imipramine might alleviate depression, even if it did not inhibit the destruction of neurotransmitters in the brain. With the problem of imipramine solved, the chemical-imbalance theory seemed to work. Two different types of drugs relieve depression, the theory went, and although they work in different ways, the net result is the same. One drug blocks the destruction of norepinephrine and serotonin. The other inhibits their reuptake. In either case, the result should be more of these neurotransmitters available in the brain.

But that was only one half of the logic behind the chemical-imbalance theory. The other half came from studies of reserpine, a drug that was extracted from *Rauvolfia serpentina* or the Indian snakeroot plant, which had historically been used to treat snakebite, hypertension, insomnia and insanity. In studies of animals, reserpine was reported to induce sedation and to decrease brain levels of norepinephrine, serotonin and dopamine. Clinical reports indicated that some people became severely depressed when taking reserpine.[14] Putting these two findings together, it seemed likely that reserpine made people depressed because it decreased neurotransmitter levels.

When the reserpine studies are added to the antidepressant studies, the logic behind the chemical-imbalance theory begins to look compelling. Drugs like reserpine that decrease monoamine neurotransmitters make people depressed. Drugs that increase these neurotransmitters by one means or another relieve their depression. Hence, depression is due to a monoamine deficiency.

THE EMPIRICAL BASIS OF THE CHEMICAL-IMBALANCE THEORY

The monoamine hypothesis made a good story. There is only one problem with it. It does not really fit the data. It didn't fit the data that were available in the 1960s when the theory was developed, and it does not fit the data that have accumulated since then.

The Myth of Reserpine-Induced Depression

Part of the initial argument for the chemical-imbalance hypothesis was that reserpine supposedly decreases the availability of monoamines and thereby makes people depressed. But does it?

Like the articles indicating that iproniazid and imipramine functioned as antidepressants, the conclusion that reserpine makes people depressed was based on clinical reports, rather than controlled trials.

In 1971 these clinical reports were carefully re-evaluated and shown to be much ado about nothing. Only 6 per cent of the people given reserpine developed clinical depression, even after taking the drug for long periods of time. This means that 94 per cent of the patients did not become depressed when given reserpine. Furthermore, of the small percentage who did become depressed, most had suffered from depression before.[15] Their new episodes of depression may have had nothing at all to do with reserpine. They may simply have relapsed, as do many depressed people after recovering from depression.[15] So much for the claim that reserpine causes depression.

The re-examination of the clinical reports showing that most people who were given reserpine did not become depressed was not published until 1971, a few years after the chemical-imbalance theory had been popularized by Schildkraut and Coppen. But a decade before their influential articles were written, there had been a carefully controlled clinical trial on the effects of reserpine on mood.[17] Far from confirming the belief that it made people depressed, the study seemed to show the reverse. Rather than making healthy people depressed, reserpine seemed to make depressed people better. As described by Michael Shepherd, the senior author of the study, in 1956:

> When we began using reserpine at the Maudsley Hospital less than two years ago there were very few reliable accounts of its use in the treatment of neuropsychiatric conditions and almost no controlled clinical studies. Dr D. L. Davies and I therefore conducted a clinical trial on a mixed group of out-patients, the majority of whom were suffering from anxiety and depressive reactions. The patients were given either reserpine, prescribed as Serpasil in a dose of 0.5 mg. by mouth twice daily, or a seemingly identical placebo, for a period of six weeks. The two substances

> were allotted by the hospital pharmacist who employed a random method and alone knew the nature of the drug dispensed to each patient; the usual crop of placebo reactions which was observed during the weekly examinations quickly demonstrated the importance of such precautions in testing patients of this type. At the end of the sixth week the response of each patient was estimated by rating scales which were completed by doctors and patients. The results demonstrated a clear-cut difference in favour of those patients treated with reserpine.[18]

In other words, the drug that was supposed to induce depression, according to the chemical-imbalance theory, actually relieved it, when it was carefully evaluated as a possible treatment in a placebo-controlled study.

How is it that the chemical-imbalance theory was proposed and so widely accepted, when the only controlled scientific study that had been done indicated that one could relieve depression, rather than induce it, by giving patients a drug that increases brain levels of monoamines? David Healy, in his comprehensive treatise on the history of antidepressants, provides an answer to this question.[19] The study was simply ignored, despite having been published in *The Lancet*, one of the world's most prestigious medical journals.

Shepherd's clinical trial of reserpine was not the only evidence that was ignored by proponents of the chemical-imbalance theory. Arguing strongly for this theory, Schildkraut and Coppen cited the work of Julius Axelrod, the Nobel Prize-winning scientist, who a few years earlier had discovered that imipramine inhibited the reuptake of norepinephrine and serotonin – the neurotransmitters that are supposed to be the causes of depression. What Schildkraut and Coppen failed to mention when arguing for their monoamine theory of depression was that Axelrod had found other drugs that inhibited the reuptake of these neurotransmitters, and one of these other drugs was reserpine – the drug that was supposed to induce depression, according to the chemical-inbalance argument.

Schildkraut and Coppen should have known that reserpine inhibited the reuptake of norepinephrine and serotonin, because it was reported in the very same articles in which the effects of imipramine had been reported. And they could not just have overlooked this detail, because it was the first thing that was mentioned in these articles. The summary at the beginning of the very first article on the subject begins with the sentence, 'Reserpine, amphetamine, imipramine, and chlorpromazine markedly reduced the uptake of circulating H^3-norepinephrine.'[20]

In a sense, it seems strange that Schildkraut and Coppen omitted this critical fact, since it might have explained Michael Shepherd's findings that reserpine functioned as an antidepressant when given to patients in a clinical trial. But acknowledging that reserpine had the same effect as imipramine on the reuptake of neurotransmitters would have demolished one of the two empirical pillars of the theory, the supposed fact that reserpine decreased levels of norepinephrine and serotonin and thereby caused depression.

What Happens When Serotonin Is Reduced

When Schildkraut introduced the monoamine theory of depression, he admitted that there was little direct evidence for it. Instead, it was based on the supposed effectiveness of antidepressant medication and the mistaken belief that reserpine makes people depressed. Schildkraut acknowledged that: 'Most of this evidence is indirect, deriving from pharmacological studies with drugs such as reserpine, amphetamine and the monoamine oxidase inhibitor antidepressants which produce affective changes.'[21] A half-century has passed since his chemical-imbalance theory of depression was introduced, and the presumed effectiveness of antidepressants remains the primary evidence in its support. But as we have seen, the therapeutic effects of antidepressants are largely due to the placebo effect, and this pretty much knocks the legs out from under the biochemical theory.

During the last 50 years researchers have tried to find more direct evidence for the monoamine theory of depression, but by and large they have failed. Instead of finding confirmation, much of the evidence they have found is contradictory or runs counter to the theory.[22] The most telling example involves techniques for rapidly reducing the amount of serotonin, norepinephrine or dopamine in the brain. The reasoning behind these studies is that if a deficiency of these neurotransmitters in the brain causes depression, then lowering their levels ought to induce depression in people who are not depressed. The evidence shows that it does not.

There are a few substances that can reduce serotonin, norepinephrine and/or dopamine rapidly and substantially, reducing them to levels thought to be lower than those of depressed patients.[23] That is what reserpine was supposed to do and, as we have seen, it did not cause depression – despite the early clinical impression that it did. Other substances have been used in later studies, the most common of which are amino-acid mixtures that lack the essential amino acids needed by the body to produce these neurotransmitters. For example, having people drink a beverage that is rich in amino acids, but does not contain tryptophan (the amino acid needed to produce serotonin), lowers their serotonin levels within a couple of hours.

Neurotransmitter depletion has been attempted in at least 90 studies and has been the subject of a number of systematic reviews, the most recent and comprehensive of which is a meta-analysis conducted by a research team at the University of Amsterdam.[24] The hypotheses of these studies were based on the premise that lowered monoamine levels cause depression, in which case depletion of these neurotransmitters ought to trigger depression in people who are not depressed. Here is what actually happens. Experimentally lowering the level of available serotonin, or of norepinephrine and dopamine, in healthy volunteers who have never been depressed does not affect their mood in the slightest.

There is only one group of research subjects in whom rapid depletion of serotonin sometimes produces clinical depression. These are depressed patients in remission who are currently taking SSRIs. About half of these patients relapse when serotonin is depleted. Note that this only happens if they are still taking antidepressant medication. If they have stopped medication, depleting serotonin may have a small transient effect on their mood, but it does not make them depressed again. And the relapse occurs only if the drug they are on is one that is supposed to enhance the availability of serotonin in the brain. Lowering serotonin levels does not cause depression in people who are currently taking a type of antidepressant that does not affect serotonin.

It is hard to know what to make of the finding that serotonin depletion depresses some people who are currently taking drugs to enhance serotonin. It could be that SSRIs induce a temporary biological vulnerability to serotonin depletion. On the other hand, it could be a 'nocebo' effect, which is a negative effect induced by a placebo. In either case, the results of decades of neurotransmitter-depletion studies point to one inescapable conclusion: low levels of serotonin, norepinephrine or dopamine do not cause depression. Here is how the authors of the most complete meta-analysis of serotonin-depletion studies summarized the data: 'Although previously the monoamine systems were considered to be responsible for the development of major depressive disorder (MDD), the available evidence to date does not support a direct causal relationship with MDD. There is no simple direct correlation of serotonin or norepinephrine levels in the brain and mood.'[25] In other words, after a half-century of research, the chemical-imbalance hypothesis as promulgated by the drug companies that manufacture SSRIs and other antidepressants is not only without clear and consistent support, but has been disproved by experimental evidence.

If the evidence for a chemical imbalance as a cause of depression is so weak, why was the theory so widely accepted and why do people still cling to it? Certainly the serotonin story was a good one – everyone could grasp it. Serotonin was good; lack of serotonin was bad. The evidence did not really fit the story, but few doctors have the time to carefully sift through the data. They see drug-company advertising in their professional journals, and they read the labelling information approved by the FDA and other regulatory agencies. At medical conferences they meet drug-company representatives, who present the company's interpretation of the evidence, an interpretation that is consistent with the simple chemical-imbalance theory. The theory may be wrong, but it certainly helps to sell antidepressant drugs, and until recently doctors have had little reason to question it.

Too Many Antidepressants Work Too Well

When the chemical-imbalance theory was introduced more than 40 years ago, the main evidence in favour of it was the contention that antidepressants, which were thought to increase the availability of serotonin and/or other neurotransmitters in the brain, seemed to be effective in the treatment of depression. As Alec Coppen wrote in 1967, 'one of the most cogent reasons for believing that there is a biochemical basis for depression or mania is the astonishing success of physical methods of treatment of these conditions.'[26] The situation has not changed very much since then. People still cite the supposed effectiveness of antidepressants as fundamental support for the chemical-imbalance hypothesis. This theory, they say, is supported by 'the indisputable therapeutic efficacy of these drugs'.[27]

Although the therapeutic effectiveness of antidepressants seemed 'astonishing' 40 years ago and still seems 'indisputable' to many people today, it is, in fact, an illusion. As I have shown earlier in this book, the difference between the effects of antidepressants and placebos is clinically insignificant, despite clinical-trial methods that ought to enhance it. But strangely enough, it is not the ineffectiveness of antidepressants that seals the fate of the chemical-imbalance theory. Rather, it is their effectiveness. The problem is that too many different types of antidepressants work too well for the theory to make physiological sense.

Different types of antidepressants are supposed to work by different means. SSRIs (selective serotonin reuptake inhibitors) are supposed to increase serotonin levels. NDRIs (norepinephrine dopamine reuptake inhibitors) are supposed to increase norepinephrine and dopamine, rather than serotonin. These two types of antidepressants are supposed to be 'selective', affecting the designated neurotransmitters without affecting the others. The strange thing is that these two types of antidepressants are equally effective in treating depression. Using data reported in a recent meta-analysis that was published in *The Lancet*, I have calculated that 60 per cent of patients respond to SSRIs and 59 per cent of patients respond to NDRIs.[28]

'So what's wrong with that?' you might ask. Maybe some people do not have enough serotonin and others do not have enough norepinephrine or dopamine. The problem is that besides the remarkable coincidence of the response rates being virtually identical, we have also accounted for too many people. Adding together 60 per cent and 59 per cent, we get 119 per cent, which is 19 per cent too much.

That may not be a big problem. It could be that some depressed people do not have a chemical imbalance and would respond to anything – even a placebo – whereas others get better only when you give them the right medication. But if this were true, then switching people to a different type of medication ought to make a difference. I reviewed the evidence on switching from one antidepressant to another in Chapter 3. Some people who have not responded to a particular antidepressant do indeed get better when you switch them to another antidepressant. The problem for the chemical-imbalance theory is that it doesn't matter what the other antidepressant is. In the STAR*D trial, which was designed to be especially representative of what happens in real-world clinical practice, switching unresponsive depressed people from one SSRI to another was exactly as effective as switching them to an NDRI. When depressed people who did not respond to an SSRI were given an NDRI, 26 per cent of them got better, but 27 per cent of them also got better if the drug they were switched to was just another SSRI.[29] Once again we have the remarkable coincidence of identical effects from different drugs.

The STAR*D trial is not alone in finding that all antidepressants are created equal. In meta-analyses of head-to-head comparisons of different antidepressants, statistically significant differences are occasionally found, but these tend to be very small

– smaller even than the clinically insignificant drug–placebo difference that we have found in our meta-analyses of the FDA data set.[30] If the difference between antidepressant and placebo is small, the differences between one antidepressant and another are virtually non-existent. As the authors of one of these analyses concluded, 'overall, second-generation antidepressants probably do not differ substantially for treatment of major depressive disorder'.[31] Furthermore, when small differences are found, they may be at least partly due to the biases in the studies. As I noted in Chapter 3, studies funded by a drug company generally report positive results for that company's drug, and negative results for drugs manufactured by competitors.[32]

The most common interpretation of the failure to find clinically meaningful differences between the effects of different antidepressants is that 'choosing the agent that is most appropriate for a given patient is difficult'.[33] This presupposes that there is a right drug for a particular patient, but the data on which this conclusion is based suggests exactly the opposite. Let us suppose that some patients have a serotonin deficiency, others a norepinephrine deficiency, and still others a shortage of both neurotransmitters in their brain. It seems a rather remarkable coincidence that the number of people suffering from all three types of imbalance would be exactly the same. But even this level of improbability underestimates how subversive the equivalence data are for the chemical-imbalance hypothesis. If some people suffer from a shortage of serotonin, others from a shortage of norepinephrine, and still others from both, then SNRIs – which are designed to increase the availability of both neurotransmitters – should provide a clinical benefit to substantially more people than either of the more selective treatments. But they do not. The effects of SNRIs are not much better than the effects of SSRIs, or than drugs like bupropion that do not affect serotonin at all.[34]

It is difficult to even imagine a convincing biochemical explanation of the virtual equivalence of different types of antidepressants. The tailoring hypothesis (the idea that the right antidepressant has to be found for each patient's particular chemical imbalance) certainly does not work. There are just too many drugs that produce response rates of 50 per cent or better in the treatment of depression, and these are not limited to antidepressants. Other drugs that work better than placebos in treating depression include sedatives, stimulants, opiates, antipsychotic drugs and the herbal remedy St John's wort.[35] I don't think anyone would argue that there is a common chemical mechanism by which all of these very different drugs work. There may indeed be different subtypes of depression, and it is plausible to suppose that different treatments might be effective for these different subtypes of the disorder. But the proportion of people having each subtype of depression cannot add up to more than 100 per cent. Yet that is exactly what the data seem to tell us, if we assume that the tailoring hypothesis is right.

Although the tailoring hypothesis does not fit the data, there is another hypothesis that works just fine. It is the idea that antidepressants are active placebos. That is, they are active drugs, complete with chemically induced side effects, but their therapeutic effects are based on the placebo effect rather than their chemical composition. Their small advantage in clinical trials derives from the production of side effects, which leads patients to realize that they have been given the active drug, thereby increasing their expectancy for improvement.

SSREs: The Last Nail in the Coffin

Different types of antidepressants are supposed to affect different neurotransmitters. Some are supposed to affect only serotonin, others are supposed to affect both serotonin and norepinephrine, and still others are supposed to affect norepinephrine and dopamine. But there is a relatively new antidepressant that has a completely different mode of action. It is a most unlikely medication, and the evidence for its effectiveness puts the last nail in the coffin of the chemical-imbalance theory of depression.

The name of this new antidepressant is tianeptine. It was developed in France, where it is licensed as an antidepressant and marketed under the name Stablon. It is also prescribed as an anti-depressant in a number of other countries, sometimes under the names Coaxil or Tatinol. Tianeptine is a selective serotonin reuptake *enhancer* (SSRE). Instead of *increasing* the amount of serotonin in the brain – as SSRIs and SNRIs are supposed to do – tianeptine *decreases* it.[36] If the monoamine imbalance theory is right, tianeptine ought to induce depression, rather than ameliorate it. But the clinical-trial data show exactly the opposite. Tianeptine is significantly more effective than placebos and as effective as SSRIs and tricyclic antidepressants.[37] In head-to-head comparisons, of tianeptine with SSRIs and with the earlier tricyclic antidepressants, all three produced virtually identical response rates. In these studies 63 per cent of patients responded to tianeptine, compared to 62 per cent of patients who responded to SSRIs and 65 per cent who responded to tricyclics.[38]

I suppose that some ingenious minds will be able to find a way of accommodating the chemical-balance hypothesis to these data, but I suspect that the accommodation will require convoluted circumventions, like those used by the Flat Earth Society in their efforts to maintain their defunct theory in the face of photographic evidence from space. If depression can be equally affected by drugs that increase serotonin, drugs that decrease it and drugs that do not affect it at all, then the benefits of these drugs cannot be due to their specific chemical activity. And if the therapeutic benefits of antidepressants are not due to their chemical composition, then the widely proffered chemical-imbalance theory of depression is without foundation. It is an accident of history produced serendipitously by the placebo effect.

Theorists Leaving a Sinking Ship

In one of the most influential books on the philosophy of science written in the 20th century, Thomas Kuhn described what happens when a prevailing scientific paradigm is on the verge of being replaced by an alternative theory.[39] The precursor to a paradigm change in science is the discovery of anomalies – findings that should not be possible if reigning conceptions were correct. As these anomalies multiply, the field is thrown into a state of crisis, from which it emerges only when the old ideas are replaced by a new paradigm.

The biochemical theory of depression is in a state of crisis. The data just do not fit the theory. The neurotransmitter depletion studies that I described earlier in this chapter show that lowering serotonin or norepinephrine levels does not make most people depressed. When administered as antidepressants, drugs that increase, decrease or have no effect on serotonin all relieve depression to about the same degree. And the effect of anti-depressants, which was the basis for proposing the chemical-imbalance theory in the first place, turns out to be largely a placebo effect.

With all of these data contradicting the chemical-imbalance hypothesis, researchers have been searching for alternative biochemical explanations. Maybe depression depends on abnormalities of the immune system, they suggest, or in a part of the brain called the hippocampus, or in the pituitary, adrenal or thyroid glands.[40] The newest fad is the theory of neural plasticity. Neural plasticity refers to the ability of the brain to change when people learn. As an explanation of antidepressant effects, the idea is that depression involves problems in the way in which depressed people process information. Antidepressant drugs – despite their very different and sometimes conflicting mechanisms – might make it easier for people to process information more efficiently and thereby learn from experience. How drugs might do this, however, is a question that has 'yet to be addressed'.[41]

The nice thing about the neural-plasticity hypothesis is that it seems to explain so much. In fact, it is a better explanation of the effects of psychotherapy than of drugs. If recovery from depression depends on learning new ways of thinking, then psychotherapy – and especially cognitive behavioural psychotherapy – ought to be effective, and indeed it is, as we shall see in Chapter 7. The concept of neural plasticity is also used to explain the therapeutic effects of electroconvulsive shock therapy, and even of placebos, on depression. As one proponent of the theory phrased it, 'psychological and pharmacological therapies, electroconvulsive shock treatment and placebo effects might all lead to improved information processing and mood recovery through mechanisms that initiate similar processes of plasticity'.[42]

The problem with the neural-plasticity hypothesis is that it does not explain how all of these very different treatments – including drugs that are supposed to have biochemical effects that are directly opposite to each other – produce their hypothesized effects on neural networks. In seeming to explain so much, the neural-plasticity hypothesis (at least as it is used as an explanation of antidepressant treatment) may actually explain nothing at all. And if placebos produce changes in neural plasticity, why bother with antidepressant drugs?

DEPRESSION, DISEASE AND THE BRAIN

As we have seen, there is no convincing evidence that depression is due to a chemical imbalance in the brain. The chemical-imbalance theory rode to fame on the basis of uncontrolled case reports of improvement on some drugs and deterioration on others, while contrary data – some of it from carefully controlled studies – were simply ignored. Later attempts to test the theory by experimentally reducing serotonin or norepinephrine in healthy volunteers disproved the theory completely. If the theory were correct, lowering the levels of these neurotransmitters in the brain ought to have induced depression. But it did not. In healthy volunteers, it had no effect at all. Finally, we were confronted with the news of the newest type of antidepressant – a drug that does exactly the opposite of what conventional antidepressants are supposed to do, and yet is just as effective as the other drugs in controlling depression. The chemical-imbalance theory is dead in the water, and its resuscitation seems an unlikely possibility.

Let me be clear. Depression certainly exists in the brain. All subjective states – sadness, joy, apprehension, delight, fear and boredom – are rooted in the brain.[43] Using sophisticated neuroimaging techniques, scientists have already established some of the neural correlates of sadness and depression, and have shown how these brain states can be altered when depressed people get better following treatment – even if the treatment is a placebo.[44] But finding that depression is represented in the brain does not mean that it is a disease, let alone a disease that can be cured by chemically altering the brain. Depression may result from a normally functioning brain, containing neural networks that have been shaped by life events and that respond to current life demands in a way that is experienced subjectively as sadness and despair. It may be the events themselves that make us feel lost and hopeless, or it may be the way in which we have learned to interpret those events. In either case, the underlying brain mechanisms may be normal.

If the chemical-imbalance theory is wrong, and if depression is not a brain disease, how is it produced and how can it be prevented and treated? One way to look for clues is to examine the process by which we were misled into the realm of chemistry. There is a culprit hiding in the history of the chemical-imbalance theory – a culprit that is guilty of leading doctors and patients astray over and over again in the history of medicine. The culprit is the placebo effect, and its darker twin, the nocebo effect. Depressed people got better when given MAO and reuptake inhibitors as antidepressants, and this led researchers to conclude that depression must be caused by a chemical deficiency. But much (if not all) of that improvement turns out to be a placebo effect. So to understand depression and how it might be treated effectively, we need to examine the placebo effect more carefully. That is the topic of the next two chapters.

5

The Placebo Effect and the Power of Belief

When our most recent – and most definitive – meta-analysis was published, the headlines in many newspapers blazoned that 'antidepressants don't work'.[1] The *Daily Telegraph* headline phrased it more specifically, clarifying that antidepressants are 'no better than dummy pills',[2] but even this headline was not entirely accurate. What our analyses actually showed was that antidepressants work *statistically* better than placebos, but that this statistical difference was not *clinically* meaningful. It was too small a difference to be of much importance in the life of a severely depressed person.

When there is little difference between a drug and a placebo, it can be due to different reasons. One possibility is that neither the drug nor the placebo is effective. A second possibility is that both are effective. When it comes to antidepressants, the latter is the case. The problem is not that people do not improve on medication. They do, and on average the degree of improvement is clinically significant. But people also improve on placebos. This suggests that it is not the drug that is making people better. Nor is it simply the passage of time or the tendency for depression to lift even without treatment. Our very first meta-analysis, in which Guy Sapirstein and I looked at the course of depression in depressed people who had not been given any treatment at all, showed clearly that untreated patients do not improve nearly as much as those given either drug or placebo treatment (see Chapter 1). So rather than being the drug or the passage of the time, it seems to be the placebo effect that makes depressed people feel better.

How can this be? How is it possible that a dummy pill with no active ingredients can produce substantial improvement in a condition as serious as clinical depression? As it turns out, placebos can be surprisingly effective, not only in the treatment of depression, but also for various other conditions. As we shall see in this chapter, placebos can reverse the effects of powerful medications. They can affect the body as well as the mind. They produce side effects as well as beneficial effects. They can make people feel sick, and they can make them feel better. Placebo effects are part of a broader phenomenon – the power of suggestion to change how people feel, how they behave, and even their physiology. If placebos can produce such powerful effects, it is important to understand them. Only by unlocking the secrets of the placebo effect can we hope to harness its power so that it can be used in clinical practice. In this chapter we look at the power of the placebo: its ability to produce therapeutic change and to cause harm.

THE DISCOVERY OF THE PLACEBO EFFECT

The word *placebo* is Latin for 'I shall please'. The medical use of the term evolved from its use in the Catholic rite of Vespers of the Office of the Dead, in which the congregation chants the words, '*placebo Domino in regione vivorum*' (I shall please the Lord in the land of the living). In medieval France, people who did not know the deceased sometimes came to funerals with the hope of sharing in the food and drinks that were distributed afterwards. These people came to be known as 'placebo singers'.

In one of the *Canterbury Tales*, Chaucer gave the name 'Placebo' to a sycophantic character, and in another (*The Parson's Tale*) he wrote that 'Flatereres been the develes chapelleyns, that syngen evere placebo' (flatterers are the Devil's chaplains, always singing Placebo).[3]

By the 19th century the term 'placebo' had entered the medical vocabulary with the meaning 'a common place method or medicine'. Still there was no recognition of the placebo effect. True to the origin of the word, placebos might please patients, but they could not make them better. The phrase 'placebo effect' does not seem to have been used prior to 1920, and the possibility that placebos might have genuine therapeutic effects was not widely recognized until the second half of the 20th century.[4] Until then, the placebo was considered a 'humble humbug' given to patients to placate them when nothing that might cure them was available.[5]

Before the middle of the 20th century clinical trials with placebo control groups were rare in medical research. Medications were adopted primarily on the basis of clinical experience and the testimony of experts in the field. A few placebo-controlled studies were done in England and the US in the early part of the 20th century, but placebos were used in these studies only as a means of controlling for the natural history of the disease – the tendency of some conditions, like the common cold, to improve without treatment. Patients in the control group were given a placebo, not because of any suspicion that the placebo might have an effect, but as a way of securing their cooperation and keeping them in the study. This was well before the age of informed consent, and medical researchers had no qualms about duping patients as part of a research project. Nowadays, deceiving patients in a clinical trial is considered unethical.

Although the routine use of placebo-controlled trials in medicine is relatively new, the logic behind it is not. Ted Kaptchuk at Harvard University has traced the use of placebo controls, although without the term 'placebo', to rites of exorcism in the 16th century.[6] The general belief was that demons could not tolerate contact with the divine and could therefore be detected and driven out by holy water, consecrated wafers and prayers. To detect fraud, priests sometimes also used 'trick trials', in which they used ordinary water instead of holy water, normal wafers instead of consecrated ones, or secular Latin texts in place of prayers. If these mundane interventions produced the same convulsions that were produced by their sacred counterparts, the diagnosis would be fraud rather than possession.

The first evaluation of a medical procedure using placebo controls occurred some 200 years later. The procedure was mesmerism – now more commonly called 'hypnosis' – as practised by the disciples of the 18th-century Viennese physician Franz Anton Mesmer. Mesmer believed that many illnesses were caused by a bodily imbalance of an invisible magnetic fluid that permeated the universe. Just as antidepressants are used today to restore a presumed chemical imbalance, so Mesmer and his followers used magnetic treatment to restore the magnetic imbalance that they believed was the cause of their patients' illnesses.

In the 18th century conventional medical treatments – like bloodletting – were not subjected to placebo-controlled trials. But mesmerism was not a conventional treatment. The mesmerists touched their patients with magnets, massaged their bodies, had them stand under 'magnetized' trees or gave them 'magnetized' water to drink. The most scandalous procedure involved having a female patient sit with her knees pressed firmly between the thighs of the mesmerist, who applied pressure to her 'ovarium', while stroking her body until she began to convulse. This was referred to as 'making passes' and, according to later historians, many women were so pleased by the convulsive crisis produced by this treatment that they followed Mesmer down the hall and begged him to repeat it.[7]

In 1784 a Royal Commission was established by Louis XVI to investigate mesmerism. Its members included some of the most illustrious figures of the time: Benjamin Franklin, who was at the time the American Ambassador to France; Antoine Lavoisier, the founder of modern chemistry; and the infamous

Joseph-Ignace Guillotin, who is now best known for his mechanical solution to the mind–body problem. The commissioners devised a series of experiments that included some surprisingly sophisticated expectancy control procedures. For example, a tree in Benjamin Franklin's garden was 'magnetized' by one of Mesmer's disciples, but the experimental subject was intentionally brought to the wrong tree. Another subject was told that a container of water had been 'magnetized'; in fact it had not. Yet another subject was misinformed that the mesmerist was 'magnetizing' her from behind a closed door. The success of these expectancy manipulations led the commissioners to conclude that the effects of mesmerism were due to imagination and belief, rather than magnetism.

Early in the 20th century German, Austrian and Swiss researchers recognized the possibility that apparent medication effects might be due to suggestion, and in the 1930s a few American medical researchers came to a similar conclusion. Still, it was not until the 1950s that the power of placebos to do more than merely provide comfort to incurable patients became widely recognized by the medical community, and it was this recognition that led to the adoption of the placebo-controlled double-blind trial – which had been advocated for two decades by Harry Gold and his colleagues at Cornell University – as the aptly named 'gold standard' for assessing new medications.[8]

THE POWER OF PLACEBO

The first influential verification of the power of placebo to produce real effects was reported in 1950 by Stewart Wolf, a physician and medical researcher at Cornell University.[9] In his seminal article on the 'pharmacology of placebos', Wolf described a number of experimental demonstrations of the ability of a placebo to reverse the effects of an active medication. In each case the reversal was brought about by misinforming the subject about the nature of the drug being administered, and in each case the subjective changes were verified by physiological assessment. The active medication was ipecac – a drug that induces nausea and vomiting and that was once used for that purpose when children accidentally swallowed a toxic substance, a practice that healthcare authorities now strongly advise against.

One of Wolf's subjects was a 28-year-old pregnant woman who had been vomiting continuously for two days. Wolf told her that he was giving her a medicine that would abolish her nausea. Instead, he gave her ipecac. Wolf reported that his patient's nausea subsided completely within 20 minutes of ingesting the ipecac syrup, and did not recur until the following morning. To see what was happening physiologically, Wolf had inserted a balloon in his patient's stomach, allowing him to record her gastric contractions. Before treatment she showed the inhibition of gastric activity that generally accompanies nausea, but when her nausea subsided, normal gastric contractions resumed. This meant that the placebo effect was not just in his patient's mind; it was also in her body.

Wolf then conducted a similar experiment on a young depressed woman who had complained of recurring episodes of nausea over the previous few months. First, he confirmed that this patient's complaints of nausea were accompanied by gastric inactivity. Then he gave her ipecac and told her it would abolish her nausea. Within half an hour Wolf observed a resumption of normal gastric activity, and the patient reported that her nausea had gone. When the nausea returned an hour later, Wolf gave her another dose of ipecac. This time the therapeutic effect occurred within 15 minutes. Normal gastric contractions resumed, and the patient reported no further experiences of nausea that day.

The best known of Wolf's demonstrations of placebo effects on gastric function involved a patient identified as Tom. Tom had a large gastric fistula, an abnormal duct that made it possible to directly observe his gastric mucous membrane. Because of his condition, Tom was the subject of more than 100 experiments on the effects of various drugs. One of these was prostigmine, a drug that produced gastric hyperfunction, abdominal cramps and diarrhoea. These effects were later reproduced by inert placebos. In another experiment, Tom was observed following 13 administrations of a placebo and during 13 control trials in which no substance was given to him. Placebo administration resulted in a 33 per cent decrease in gastric acid secretion, as compared to an 18 per cent decrease during control trials.[10]

In 1955 Henry Beecher published an article entitled 'The Powerful Placebo', which, despite its age, may be the single most influential paper on the placebo effect ever written.[11] Beecher claimed that, averaged across 15 studies involving a variety of conditions – including severe post-operative pain, headache, anxiety, seasickness, coughs and colds – about one out of three patients given a placebo showed significant improvement, a figure that has come to be enshrined as gospel. Yet as widespread as this conventional wisdom is, it is a myth.

In fact, the percentage of patients who respond to a placebo can vary from none at all to almost everyone. That the response to a placebo can vary widely was first shown in a 1957 study conducted by Eugene Traut and Edwin Passarelli at an arthritis clinic in Chicago.[12] First, Traut and Passarelli gave their patients placebo pills. Half of the patients improved; half did not. Those showing no improvement were then given placebo injections. Adding together those who responded to the placebo pill and those who responded to the placebo injection, 82 per cent of the patients seemed to benefit from placebo treatment, and continued placebo treatment was effective for up to 30 months, which was the full duration of the study. So what is the real rate of response to a placebo? Is it 30 per cent, as Beecher had claimed; 50 per cent, as shown by patients given placebo pills for arthritis; or 80 per cent, as shown when the pills were supplemented by placebo injections? In fact, the question does not have a meaningful answer. It is much like asking what percentage of people get drunk on beer, without specifying how much beer they have consumed.

Although Beecher's paper on the power of placebo was enormously influential – to the point of changing the way new medicines are evaluated – there was a fundamental flaw in the data upon which it was based.[13] Sometimes people improve without being given any treatment at all, not even placebo treatment. Beecher's estimate of the rate of response to a placebo did not take into account the natural history of the condition being treated, spontaneous recovery or any of the other factors that can produce improvement even in patients who have not been given a placebo. Certainly the 35 per cent placebo response that Beecher calculated for the common cold must have been due to the simple passage of time. Just as the effect of a drug is assumed to be the difference between the response to the drug and the response to a placebo, so the placebo effect is the difference between the response to the placebo and improvement that would have occurred if the person had not taken a placebo – and that is something Beecher simply did not evaluate.

In 2001 two Danish researchers, Asbjørn Hróbjartsson and Peter Gøtzsche, published an influential meta-analysis in which they estimated the difference between the effects of getting a placebo versus doing nothing at all.[14] Although they found a significant placebo effect, especially in the treatment of pain, the overall effect seemed very small – much smaller than would have been expected of a 'powerful' treatment. On the basis of these data, the researchers asked 'Is the placebo powerless?' and answered their own question by concluding that there was little evidence that placebos have powerful clinical effects.

It seemed that Beecher was wrong after all. But was he? There are two major problems with the Danish meta-analysis. One problem is the way in which the term 'placebo' was defined. Usually, placebos are dummy pills, capsules or injections, presented in the guise of active medications. But many of the studies that Hróbjartsson and Gøtzsche evaluated did not include a placebo in this sense of the term. Instead, these studies looked at the effects of leisure reading, answering questions about hobbies, and talking about books, movies and television shows. All of these were called placebos, and their effects were included as placebo effects. But do they really qualify as placebos? If you were given a medication and told by your doctor that it had been proven effective, you might have considerable confidence in it. But imagine that instead of giving you medication, your physician asked you about your favourite television programme or suggested that you curl up with a nice mystery that night. How likely is it that you would go away with the expectation of improvement that placebos are supposed to generate, and by means of which they are presumed to produce their effect? A meaningful evaluation of the placebo effect has to be based on a credible placebo, one that raises expectations of improvement that are as great as those elicited by active treatment.

An even more fundamental shortcoming in Hróbjartsson and Gøtzsche's analysis is the diversity of disorders that they evaluated. These included the use of placebos to treat the common cold, infertility, marital discord, mental retardation, alcohol abuse, smoking, poor oral hygiene, herpes-simplex infection, fecal soiling and 'undiagnosed ailments'. Placebos are not panaceas. They may be very powerful for some conditions, less effective for others, and have no effect at all on some ailments. As we saw earlier in this book, placebos are highly effective in the treatment of depression, in which the placebo effect (that is, the difference between the response to the placebo and the mere passage of time) is twice as large as the drug effect (the difference between the response to the drug and the response to the placebo). They also have a substantial effect on pain, especially in studies specifically designed to assess the placebo effect.[15] But placebos are not likely to have much of an effect on infertility. Nor are they likely to have any effect on newborn infants, who were the subjects in one of the studies that Hróbjartsson and Gøtzsche analysed.

Imagine reading a scientific article assessing the effectiveness of medical treatment in general, without regard to what condition was being treated or how it was treated? The article might conclude that medical treatments are very effective or that they are not very effective, depending at least in part on which particular medical treatments had been included in the review. It is just not meaningful to try and estimate the effectiveness of medical treatment in general. Some medical treatments are extremely effective, whereas others have much smaller effects, and there are some medical conditions for which effective treatments have not yet been found.

This is the basic problem with any attempt to evaluate the overall effectiveness of placebos, as Beecher and the Danish researchers had tried to do. There is not just one placebo effect. Instead, the placebo effect depends on a host of factors. It depends, for example, on the condition being treated, the way in which the placebo is administered, the colour of the placebo, its price, whether it has a recognized brand name and the dose that is prescribed. Studies of the placebo effect reveal that, all else being equal, taking placebo pills four times per day is more effective than taking them only twice a day; brand-name placebos are more effective than placebos presented as generic drugs; placebo injections are more effective than placebo pills; and more expensive placebos are better than cheaper ones.[16]

The placebo effect also depends on what people are told about the 'treatment' they are given. The effect is smaller when patients are told that their treatment might be a placebo, as is routinely done in clinical trials, and is larger when people are told that their treatment has been shown to be powerful.[17] Because the placebo effect can vary so much, attempts to estimate its power in general, without specifying the condition being treated and the conditions under which the placebo was administered, are meaningless.

Perhaps the most persuasive evidence that the placebo effect can be very powerful comes from studies in which it has been found to be more effective than an active drug. The most recent study of this sort was reported by a team of researchers led by Peter Tyrer in the Department of Psychological Medicine at Imperial College London. Tyrer's group assessed the use of antipsychotic drugs as a way of reducing aggressive behaviour in mentally retarded adults. Consistent with clinical reports, they did indeed find a substantial drop in aggression following treatment, but the largest decrease in aggression was not in the groups treated with real drugs. Rather, it was in those given a placebo. The patients given the real antipsychotic drugs showed an average decrease in aggressive behaviour of about 60 per cent. Those given a placebo showed an 80 per cent reduction in aggression. It seems that the effect of the real drugs was to reduce the powerful placebo effect.[18]

Placebo Surgery

One of the factors that influence the magnitude of the placebo effect is the way in which the placebo is administered. Placebo injections, for example, are more effective than placebo pills; and placebo acupuncture – which uses sham needles that retract into their handles like the blade of a stage dagger, rather than piercing the skin – is also more effective than placebo pills.[19] The most powerful placebo of all is surgery. Approximately 45 per cent of patients with Parkinson's disease get better when treated with sham surgery, but only 14 per cent of Parkinson's disease patients improve when treated with placebo pills.[20]

Placebo surgery? I know it sounds like a joke, but it isn't. Like any medical treatment, surgical procedures elicit expectancies of improvement, and therefore part of their effectiveness can be due to the placebo effect. For this reason, sham surgery has been used as a placebo in some clinical trials. Placebo surgery consists of cutting the patients open and sewing them up, without doing the actual surgical intervention.

The first studies using placebo surgery were done at the end of the 1950s. At that time, a surgical procedure called mammary ligation was used to treat angina pectoris. Angina is a chest pain that occurs when the heart muscle does not get enough oxygen. It is a symptom of coronary artery disease, which is produced when plaque narrows or blocks the arteries, thereby reducing the flow of oxygen-rich blood to the heart. The theory behind mammary ligation was that if some of the coronary arteries were blocked off completely, the blood would find alternative routes to the heart.

Clinical experience indicated that mammary ligation was very effective in the treatment of angina, with success rates as high as 85 per cent.[21] Still, at least part of its effectiveness might be due to the placebo effect, the power of which had recently been promulgated in the articles by Beecher, Wolf and Gold that I described earlier in this chapter. This possibility led two independent teams of medical investigators, one in Seattle and the other in Kansas City, to test the effects of mammary ligation against placebo surgery.[22] Some patients in these studies were given the real surgical treatment. Patients in the placebo groups were also cut open and their mammary arteries were exposed, but the arteries were not tied off. Across these two clinical trials, 73 per cent of the patients receiving real mammary ligation showed substantial improvement. This is not much different from what had been reported in uncontrolled studies, and had the researchers not included placebo surgery groups, they might have concluded that mammary ligation was effective. But the rate of improvement with placebo surgery was 83 per cent, which was not significantly different from the response to the real treatment. The apparent effect of mammary ligation was gigantic, larger than the effect of giving antidepressants to depressed patients, but it was a placebo effect. Needless to say, mammary ligation is no longer used as a treatment for angina.

The patients' comments following placebo surgery is instructive. Asked whether they had noticed any change following surgery, one patient said, 'Yes. Practically immediately I felt better. I felt I could take a deep breath and I have taken about ten nitroglycerins since surgery. These pains were light and brought on by walking. I figure I'm about 95 per cent better. I was taking five nitros a day before surgery. In the first five weeks following, I have taken a total of twelve.' Another patient, responding three months after the operation, wrote, 'I can do anything except real hard lifting. I am running farm equipment and using maybe one nitro a week. I used to need fifteen a day. Believe I'm cured.'[23]

These comments are very similar to testimonials for antidepressants that appeared in the media shortly after my most recent meta-analysis was published. They demonstrate the danger of relying on clinical reports of patient improvement in deciding whether a particular treatment is effective. Placebos can yield substantial clinical benefit that can last for months or even years.[24]

After these two placebo-controlled clinical trials of mammary ligation, the use of a placebo control condition to evaluate surgical procedures seemed to disappear, only to make a comeback some 40 years later. In the 1990s Bruce Moseley, a surgeon at the Veterans Affairs Medical Center in Houston, Texas, and physician for the Houston Rockets basketball team, routinely performed arthroscopic surgery for osteoarthritis of the knee. Two procedures were in use at the time, and there was a debate as to which was better. One procedure involved making small incisions in the knee and rinsing the joint. In the second procedure, the joint was scraped as well as rinsed. Some doctors thought that scraping rough surfaces of the joints made the operation more effective, whereas others suspected that it might cause some damage.

Although these were well-established operations, performed on hundreds of thousands of people each year, Moseley wondered whether either procedure was of any real benefit, and conceived the idea of comparing them directly. When Moseley proposed this idea to Nelda Wray, his colleague at the VA hospital and Director of Health Services Research at the Baylor College of Medicine, she suggested that the apparent benefits might be due to the placebo effect. At first, Moseley was sceptical. This was a surgical procedure, after all, not a sugar pill. But Wray convinced him that the possibility was worth investigating. 'The bigger and more dramatic the patient perceives the intervention to be,' she said, 'the bigger the placebo effect.'[25]

Wray and Moseley then assembled a team of medical researchers and designed a clinical trial aimed at comparing real arthroscopic surgery to placebo surgery.[26] They recruited 180 patients for the study. One-third of them were given the full rinsing and scraping procedure. For another third of the patients, the joint was rinsed, but not scraped. The rest were given placebo surgery. First, three incisions were made with a scalpel so that there would be scars afterwards. Then 'the surgeon asked for all instruments and manipulated the knee as if arthroscopy were being performed. Saline was splashed to simulate the sounds of lavage.'

Not only was this placebo operation effective, but it was significantly more effective than actual surgery, at least in the short run. Two weeks after their operations, patients in the placebo group reported significantly less pain than those in either of the surgery groups, and they also showed more improvement on an objective test of walking and climbing stairs. One year after the operation, patients in the placebo group still walked and climbed stairs significantly better than those whose knee joints had been both rinsed and scraped, and two years after the surgery there were no significant differences between the groups. In other words, in the long run, rinsing the knee joint did no good at all, and – as Moseley had expected – scraping it actually caused damage lasting at least a year.

There are some interesting parallels between Moseley and Wray's study of arthroscopic surgery and the meta-analyses that my colleagues and I reported for antidepressants. One similarity is that the failure to find substantial differences between real and placebo treatment was not because of a lack of response to the treatment. Patients given real surgery in Moseley and Wray's study reported having much less pain than they had before treatment, just as patients given antidepressants report being less depressed. But in both cases, patients also showed substantial improvement after placebo treatment. One patient in the sham-surgery group described the outcome as follows: 'The surgery was two years ago and the knee never has bothered me since. It's just like my other knee now. I give a whole lot of credit to Dr Moseley. Whenever I see him on the TV during a basketball game, I call the wife in and say, "Hey, there's the doctor that fixed my knee!"'[27]

Another parallel between Moseley and Wray's study of sham surgery and the study in which my colleagues and I reported our analysis of the FDA antidepressant data is the reactions that they evoked. If real arthroscopic knee surgery is no better than placebo surgery, one would think that the procedure would be abandoned, just as mammary ligation was discarded as a surgical procedure after its effects were found to be no better than those of placebo surgery. Instead, many orthopaedic surgeons tried to discredit Moseley and Wray's clinical trial, just as defenders of antidepressants have tried to discredit our meta-analyses of antidepressant drugs. An editorial in the journal that published the Moseley and Wray study, for example, opined that arthroscopic surgery might benefit some patients but not others.[28] In a spirited and compelling reply to the editorial, the authors of the study responded that 'if someone questions whether arthroscopic surgery would be efficacious in a specific subpopulation of patients, then the ethical way to proceed would be to test the hypothesis by conducting a placebo-controlled trial in that specific subgroup'.[29] I agree completely, and the same can be said for those who hypothesize that antidepressants might be clinically effective for some subgroups of patients. If antidepressants are effective for some groups of patients, 'the ethical way to proceed would be to test the hypothesis by conducting a placebo-controlled trial in that specific subgroup'.

More recently, new data have confirmed the findings reported by Moseley and Wray. The new study showed that surgery added nothing to the effects of physical and medical therapy alone.[30] This article was also accompanied by an editorial offering a defence of arthroscopic surgery.[31] The editorial acknowledged that the new study provided 'strong support for the conclusion of Moseley et al. that arthroscopic surgery is not effective therapy for advanced osteoarthritis of the knee', but added that perhaps it is useful for patients whose osteoarthritis is accompanied by some other knee condition. It's déjà vu all over again!

Is placebo surgery ethical? Should doctors be allowed to administer anaesthesia and make surgical incisions, but then not do any therapeutic intervention? Isn't the first rule of medicine to do no harm? It is true that informed consent is now required in clinical trials. This means that patients are told that they might not get the real surgery, and they can decide whether or not to participate. But is that enough? Is it acceptable to expose patients to the risks of sham surgery, even if they agree to participate?

This is surely an ethical dilemma, but there is also another way to look at this issue. The question can be rephrased as follows: is it ethical to perform a surgical procedure on patients without first testing it against placebo surgery? Suppose that the placebo-controlled studies of mammary ligation had never been done at all. We would never have learned that the benefits of this surgical procedure are due to the placebo effect, and we would still be performing this ineffective procedure today. Over the years it would have been performed on hundreds of thousands of patients, without the patients or their physicians ever knowing that the surgery was really a placebo. So the choice is between giving sham surgery to a relatively small number of patients, after informing them of the risks and letting them decide whether to participate, or exposing large numbers of patients to the risks of ineffective surgery, with neither them nor their doctors knowing that the surgical procedure is, in fact, merely a placebo. Which of these alternatives do you consider ethically preferable?

Mind and Brain

One of the factors that determine the effectiveness of a placebo is the nature of the condition being treated. Conditions that have a strong psychological component – such as pain, anxiety and depression – are particularly prone to placebo influence, whereas conditions like bone fractures, diabetes and infertility are less likely to be affected by placebo treatment. But this does not mean that placebo effects are 'all in the mind'. Placebos affect physiology as well as psychology.

The most common physiological effects of placebos are those that are associated with changes in subjective experience. When placebo stimulants make people feel energized and alert, for example, they also increase their blood pressure and heart rate, and when placebo tranquillizers relax people, they decrease their blood pressure and heart rate.[32] Similarly, when Stewart Wolf gave ipecac to patients and told them it would ease their nausea, their reports of no longer feeling nauseous were accompanied by a resumption of normal gastric activity.

Many people seem particularly impressed by the physiological effects of placebos. They see them as evidence that the mind can affect the body. But the physiological placebo effects I have described are not all that surprising. Instead, they are exactly what we should expect, given what we know about the relation between mind and body. Strictly speaking, they are not really instances of mind affecting body. Rather, they are instances of body affecting body.[33]

What do I mean by this seemingly strange assertion? As far as we know, there is a physical substrate to all of our subjective experiences. In particular, experience seems to be linked to our brains. When the brain is injured, subjective experience is also changed, and the changes in experience are specific to the location of the tissue damage. Conversely, our subjective experiences are accompanied by changes in brain activity, and the particular areas of the brain in which these changes occur depend on the nature of the experience. With the advent of modern methods of imaging the brain, neuroscientists have located specific brain areas that are involved in vision, pain perception, speech, the voluntary control of movements and a vast myriad of other cognitive functions that were in the past attributed to the mind. Just as water is H_2O, so the mind seems to be the brain.[34]

If what we experience is associated with something that happens in the brain, and if placebos change subjective experience, then we ought to be able to find changes in brain activity that are produced by placebos – and in fact this is precisely what has been found. A team of researchers led by Helen Mayberg, a neurologist at Emory University and the University of Toronto, have used a technique called positron emission topography (PET) to study changes in brain activity associated with the experience of depression.[35]

In the first of these studies, the researchers identified the areas of the brain that are associated with normal sadness. They asked volunteer subjects to think about some very sad personal experiences – and about some emotionally neutral experiences – while their brains were being imaged in a PET scanner. When thinking about the sad experiences, the volunteers reported feeling intense sadness, and many of them became tearful. The PET scans showed the changes in brain activity that accompanied these sad feelings. They demonstrated increased blood flow in the limbic system – a part of the brain that is involved in the control of emotion – and decreased blood flow in parts of the brain that are involved in the control of attention.

In their next study, Mayberg and her colleagues scanned the brains of depressed patients who had responded positively to treatment for depression in a clinical trial of Prozac. The patients were scanned twice, once before the treatment had begun and once again after six weeks of treatment. About half of the patients responded positively to the treatment by showing at least a 50 per cent reduction in their symptoms; the other half did not improve that much and were classified as non-responders. For the responders, but not for the non-responders, treatment of depression produced changes in brain activation in exactly the same areas in which normal sadness had produced changes, but in the opposite direction. In other words, successful treatment decreased brain activity in areas where sadness produces increased activity, and it increased brain activity in areas where sadness decreases it.

At first blush, you might be tempted to interpret this as evidence for a specific neurophysiological effect of Prozac on depression. But there was a catch. Only half of the successfully treated patients had been given Prozac. The rest had recovered on a placebo, and the changes in brain activity that the researchers had found were 'independent of whether the substance administered was active fluoxetine or placebo'.[36] In other words, when placebos are successful in lowering depression, they also produce changes in brain activation, and for the most part these are the same changes produced by the real drugs.

Mayberg's studies seem to suggest two conclusions. First, one would be tempted to conclude that it is the placebo effect, rather than the chemical effect of medication, that had changed the brain activity of the patients who had been given Prozac. Second, one might interpret the observed changes in brain activity as indications of how placebos reduce depression. In fact, neither of these conclusions is justified. The physiological changes are exactly what would be expected of *any* effective treatment for depression, no matter how the treatment works. They are changes in patterns of brain activity that correspond to sadness and depression. When depression is overcome, these changes in brain activation are reversed, no matter how the improvement in depression is brought about, whether by drugs, placebos or some other form of treatment.

Each of these treatments might also produce physiological alterations that are specific to the treatment.[37] Antidepressants are active drugs, and like other active drugs they certainly have physiological effects. Psychotherapy is a learning experience, and learning changes the brain.[38] So it, too, has specific neurological effects. Nevertheless, recovery from depression has its own neural substrate, and this can be seen when people improve on placebos, as well as when other treatments have produced the improvement.

Depression is not the only clinical condition in which placebo effects have been linked to changes in the brain. Changes in brain activity have also been shown in neuroimaging studies of placebo analgesics, the most influential of which was reported by a team of researchers led by Tor Wager, a neuroscientist at Columbia University who, at the time he conducted these studies, was a postgraduate student at the University of Michigan.[39]

Wager's interest in the connection between mind and body stemmed from childhood. He had developed a severe skin rash, and his mother – who was a Christian Scientist who believed that illnesses were products of the mind – prayed and prayed for its cure, but to no avail. Finally a friend said to her, 'Enough praying; take the kid to a doctor.' The doctor applied an ointment to the skin and the rash was cured. From that point on, the family took a more traditional approach to medical treatment.

With that story as part of his family lore, Wager grew up sceptical of claims about the healing power of the mind, but he also developed a keen interest in the kinds of health-related outcomes that might be affected by belief. That interest eventually led him to conduct scientific investigations of the placebo effect, studies that have given him an international reputation.

Wager's first step was to see if giving people a placebo would lead them to report less pain. Despite what he had read about the placebo effect, he was not convinced that it would. To find out, he and his colleagues induced pain in healthy volunteers with electric shocks. Sometimes the investigators put a placebo cream, which they described to subjects as 'a highly effective pain-relieving medication', on the subjects' arms before shocking them. Sometimes they shocked them without the cream. In either case, the subjects had to rate how much pain they felt.

The volunteers in Wager's study reported feeling less pain when the placebo cream had been applied than when it had not. In other words, they showed a placebo effect. But had they really felt less pain, or was this just something they were saying to be cooperative? To answer this question, Wager ran two more studies. They were much like the first, but this time the experimenters induced pain – with and without placebo treatment – while imaging the subjects' brains in an fMRI (functional magnetic resonance imaging) scanner.

When the placebo cream had not been applied, the researchers found activation in areas of the brain that they identified as the 'pain matrix'. But when the same pain stimuli were administered with the placebo cream, activation in these pain-responsive regions of the brain was reduced, and the more pain relief the subjects reported, the greater the reduction of activation in the pain matrix. This told Wager that people actually do experience less pain when given placebo analgesics, and this change in experience is accompanied by changes in brain activity.

Brain and Body

The physiological effects that we have reviewed so far – be they changes in heart rate, blood pressure or brain activation – are exactly what one would expect, given that any change in experience should be associated with a corresponding change in physiology. But other placebo-induced physiological effects have been reported in the literature, and some of these are more difficult to understand. These include physiological effects of placebo treatment on asthma and eczema. If we are right in assuming that the mind is the brain, then they are really examples of the brain affecting other parts of the body, but exactly how it does so in each instance remains unclear.

The placebo effect in asthma is one of the most well-studied and robust placebo effects on physiological function. The wheezing that sufferers of asthma experience is accompanied by a constriction of the bronchial airways that makes it difficult for them to breathe. Asthma medications dilate the bronchial tubes, making it easier to breathe, but a large number of studies have shown that placebos can also affect bronchial dilation. In fact, about two-thirds of the response to real asthma medication is also produced by placebo treatment, leaving about one-third of the effect as a true drug effect.[40]

One of the most convincing demonstrations of the effect of placebos on asthma was conducted by a research team led by Thomas Luparello, a psychiatrist at the State University of New York.[41] Luparello's team asked 40 asthmatic patients to inhale what they presented as irritants or allergens previously identified by the subjects as triggers for their asthmatic attacks. In fact, the substance they inhaled was an inert saline solution – simple table salt dissolved in water. Nineteen of the 40 asthmatic patients reacted with a significant increase in airway resistance, and 12 of them developed full-blown bronchospasm attacks. These asthma attacks were then reversed by the administration of a placebo presented as an asthma medication.

In a subsequent study, Luparello and his colleagues gave two different drugs to asthmatic patients.[42] One of the drugs was a bronchodilator, a medication that dilates the airways and makes it easier to breathe. The other was a bronchoconstrictor, a drug that constricts the bronchial tubes and makes breathing more difficult. Sometimes the subjects were told the truth about what they were inhaling. Sometimes they were misled – they were told they were inhaling a bronchodilator when in fact they were inhaling a bronchoconstrictor, or vice versa. After each inhalation, the researchers measured changes in airway resistance. Not surprisingly, they found a significant effect for the type of drug the patient inhaled. The bronchodilator dilated the bronchial tubes and the bronchoconstrictor constricted them. But what the subjects were told also made a difference. When the suggestion about the effect of the drug was in conflict with the real drug effect, the response to the drug was cut in half.

In 1962, Yujiro Ikemi, founder of the Japanese Society of Psychosomatic Medicine, and his colleague Shunji Nakagawa published a remarkable study showing that suggestion could both induce and inhibit contact dermatitis.[43] Contact dermatitis is a skin condition produced by chemical substances to which people have become sensitized. One of these substances is an oil called urushiol, which is found in various plants, including poison ivy in the United States and lacquer trees in Japan. Some people are very sensitive to urushiol; others much less so. Ikemi and Nakagawa found 13 boys who reported being hypersensitive to lacquer leaves. They touched one of each boys' arms with leaves from a harmless tree, telling them that the leaves were from a lacquer tree. On the other arm the students were touched with the poisonous lacquer leaves, which they were told were from a harmless chestnut tree. All 13 boys displayed a skin reaction to the harmless leaves (the placebo) and in 11 of these boys the reaction was described as 'marked'. Only two of the boys reacted to the actual poisonous leaves.

Perhaps the most provocative report of placebo power is a case in which placebo treatment appeared to have profound effects on the course of cancer.[44] Mr Wright', as he was called in the report of his case, had tumours the size of oranges in his neck, armpits, groin, chest and abdomen. The prognosis was that he had less than two weeks to live. Having read about a new experimental drug that was to be tested at the hospital, Mr Wright persuaded his physician to include him in the clinical trial. Three days later the tumours had 'melted like snow balls on a hot stove, and in only these few days, they were half their original size'. Within ten days practically all signs of the disease had vanished.

About two months later reports began appearing in the press indicating that the experimental drug had been proven ineffective. After reading these reports, Mr Wright lost faith in the treatment that seemed to have benefited him so greatly and relapsed to his pre-treatment condition. At this point, his physician managed to persuade him that the negative results were due to a deterioration of the drug and that a new, refined, double-strength product was due to arrive shortly. A couple of days later, treatment with an inert placebo was begun.

The effects of placebo treatment were even more dramatic than those obtained with the experimental drug. Once again the tumour masses 'melted away' and Mr Wright remained symptom-free for two more months. Then he read an announcement by the American Medical Association concluding that the drug he thought he was getting was 'worthless'. He died a few days later.

As provocative as it is, we have to be careful in drawing conclusions from Mr Wright's case history. At best, it is a tantalizing teaser. It is, after all, based on only one patient, and it is most likely that the changes in Mr Wright's condition were not due to his belief in the medication. There is evidence that some cancers may spontaneously go into remission,[45] and this might be the best explanation of reported changes in Mr Wright's cancer. The timing of the changes in his condition – the fact that remission occurred when he thought he was taking an effective medication and that he relapsed when he learned that the medication was ineffective – might just have been coincidental. As convinced as I am by the data that there is a powerful placebo effect on some conditions, I remain sceptical of claims of remarkable cures of physical illnesses.

As Carl Sagan said, 'Extraordinary claims require extraordinary evidence'. With respect to seemingly miraculous cures of serious physical conditions, even ordinary evidence is largely missing. Nevertheless, Mr Wright's story seems compelling enough to suggest that controlled research on the ability to affect cancer psychologically might be warranted.

The Nocebo Effect

We usually think of placebo effects as being beneficial. Placebos reduce depression, anxiety and pain, improve the symptoms of Parkinson's disease and open up the constricted airways of people suffering from asthma. But placebos can have negative as well as positive effects, a phenomenon that is called the nocebo effect. We have already encountered some of these. Just as placebo inhalants can open airways, they can also constrict them. It all depends on what the person is told about the substance they are inhaling. In the Japanese study on contact dermatitis, being touched with placebo leaves produced skin reactions. In the case report of Mr Wright's placebo treatment for cancer, although the patient went into remission when he thought he was getting effective treatment, he relapsed when he became convinced that the treatment was ineffective. All of these are examples of the nocebo effect.

One of the most fascinating examples of the nocebo effect comes from a study of the effect of placebos on insomnia.[46] Two researchers at Yale University advertised for students suffering from insomnia to participate in a study that was supposedly investigating the effect of bodily activity on the content of their dreams. Some of the subjects were given placebo pills to take before going to sleep; others were in a control group and were not given any pills. Students on the pills were given different information about what the pills contained. Half of them were told that the pill would arouse them; the other half were told that it would relax them. The results were surprising. Insomniac students given the 'arousing' pills fell asleep sooner, and those given the 'relaxing' pills took longer to fall asleep.

How did the researchers explain these strange findings? Actually, they had predicted the results in advance. Their idea was that when people have trouble falling asleep, they may see the cause of their difficulty as being a personal inadequacy. This attribution about why they cannot sleep makes the person emotionally aroused, thereby making it even harder to fall asleep. When given 'arousing' pills, however, people may make a different interpretation about the meaning of their arousal. Now it isn't an indication of a personal characteristic, but rather a condition produced by the drug. This leads them to worry less about their sleeplessness and therefore to fall asleep more easily. On the other hand, when people given 'relaxing' pills find themselves having difficulty settling down, the fact that they have taken a pill might intensify their negative attributions. 'Look how badly off I am,' they might think. 'I've taken a tranquillizer and I *still* can't sleep. I must really be in bad shape.' Thoughts like this would, of course, make it even more difficult to get to sleep.

Sometimes nocebo effects spread like infectious diseases and affect large numbers of people. Technically, this is called mass psychogenic illness, but it is more commonly known as mass hysteria. Mass hysteria has been recognized for centuries, but a relatively recent case that was reported in the *New England Journal of Medicine* a few years ago provides a nice illustration of the phenomenon.[47] On 12 November 1998 a high-school teacher in the state of Tennessee noted a smell like that of petrol in the classroom, following which she reported experiencing a headache, nausea, shortness of breath and dizziness. When some of her students reported similar symptoms, the class was evacuated, the school was closed and 100 students, staff and family members were taken to the emergency room of the local hospital, where more than one-third were kept overnight.

The school remained closed for two days, during which it was examined carefully by the fire department, the gas company and state officials of the Occupational Safety and Health Administration (OSHA), but no evidence of any toxic compounds was found. Meanwhile, the number of students and staff experiencing symptoms increased, and the variety of symptoms they reported expanded to include tightness of the chest, difficulty breathing, sore throat, burning eyes, coughing, abdominal pain, watery eyes, vomiting, sneezing and diarrhoea, but no blood or urine abnormalities were detected. Only three factors predicted whether a student developed symptoms. Students were more likely to have symptoms if they were female, if they had seen someone else showing symptoms and if they knew someone who had developed symptoms. The investigators concluded that the case was typical of mass hysteria.

Inspired by the Tennessee school incident, William Lorber, Giuliana Mazzoni and I have studied psychogenic illness in the laboratory.[48] We asked a group of university students to inhale a substance that we described to them as a suspected environmental toxin. In fact, what we gave them to inhale was plain ambient air. We told them that the substance had been reported to evoke a number of physical symptoms, particularly headaches, nausea, drowsiness and itchy skin. Then we had them report their experience of these and other symptoms over the course of an hour. During that time the students reported an increase in all of the symptoms. The increase was much larger for the four symptoms that we had identified as having been reported previously, and their reports of itchy skin and drowsiness were accompanied by scratching and yawning. We carried out this study in the state of Connecticut in New England, where polluted air might indeed contain toxic substances, but we also included a control group. The students in the control group were not asked to inhale from the placebo inhaler, but of course they were breathing the same air. They did not report an increase in physical symptoms.

Throughout this book I have stressed the importance of side effects in clinical trials of antidepressants. They can tip patients off to the fact that they have been given the real drug rather than a placebo, leading them to both expect and experience greater improvement than patients who have been given placebos. Sometimes placebos also produce side effects, especially those that the patients expect. This was first demonstrated in a study of headaches as a side effect of lumbar puncture, a clinical procedure used to administer anaesthetics or to extract spinal fluid for diagnostic purposes. The researchers performed lumbar punctures on two groups of patients, but only warned one of them that headaches were a possible side effect. Approximately half of the subjects who were forewarned subsequently reported headaches, as compared to only one of the control subjects, suggesting that this commonly reported consequence of lumbar puncture may be a nocebo effect.[49]

A similar result was reported in a multi-centre clinical trial of aspirin as a treatment for angina. Two of the three institutions participating in the study listed gastric irritation in their list of possible side effects; the third centre did not. The results of the study showed that significantly more patients reported gastrointestinal symptoms in the institutions that had listed gastric irritation on the informed-consent from than in the centre that had made no mention of this possibility.[50] The results of these studies present doctors with an ethical dilemma. On one hand, we have the requirement for informed consent, according to which we should warn people in advance of the possible side effects they might experience. On the other hand, warning them may produce side effects that would not otherwise have occurred. So what should we do? I really do not have a solution to this dilemma, but it is certainly a problem that deserves – but has not as yet received – careful consideration.

In a particularly dramatic case of placebo-induced side effects, doctors at a hospital in Jackson, Mississippi, treated a young man who came into the emergency room, said to the receptionist, 'Help me, I took all my pills' and then collapsed to the floor, dropping an empty prescription container. His blood pressure was abnormally low, and he was treated with intravenous fluids, which brought it back to within a normal range. The prescription bottle bore a label indicating that the medication was part of a clinical trial of antidepressants. Further investigation revealed that he had been assigned to the placebo group. He had overdosed on a placebo.[51]

Even more dramatic are reports of death by placebo, although they remain controversial. In 1942, Walter Bradford Cannon, a distinguished physiologist and chair of the Department of Physiology at the Harvard Medical School, wrote an article entitled '"Voodoo" Death', in which he recounted numerous instances in which people in tribal societies were reported to have died after having been cursed or having violated strict taboos.[52] Cannon offered the explanation that voodoo death, if real, might be caused by intense fear. The victims literally died of fright. I cannot say I am convinced that voodoo death occurs. I am sceptical about its reality, primarily because the evidence for it is anecdotal, but Cannon's physiological explanation of how the fear of death might cause a person to die remains valid. Writing on the 60th anniversary of the article's publication, Esther Sternberg, Director of the Integrative Neural Immune Program at the National Institute of Mental Health in the US, concluded that it was remarkably accurate and had withstood the test of time.[53]

Does the production of side effects by placebo undermine my argument that the perception of these effects can lead patients to realize that they have been given the real drug, thereby producing an enhanced placebo effect? Not really. Although placebos can induce side effects, antidepressants produce significantly more of them. In one clinical trial, for example, 19 per cent of the patients given a placebo reported adverse events, but 46 per cent reported side effects on an antidepressant, and as I mentioned in Chapter 1, once you adjust for drug-placebo differences in side effects, the difference in therapeutic benefits is no longer significant, not even statistically.[54] Some of the adverse events that patients report may not be side effects at all. They might have occurred even if the person had not been treated. That is why they are often called 'adverse events' rather than 'side effects'. But the difference in adverse events between drug groups and placebo groups is most certainly an indication of drug-induced side effects, and can easily explain the small drug-placebo difference in improvement.

DEPRESSION AS A NOCEBO EFFECT

As I will discuss in greater detail in the next chapter, placebo and nocebo effects are part of a broader phenomenon – the tendency for people to experience what they expect to experience.[55] Is it possible that negative expectancies can make people depressed? If so, it would help explain the powerful effect of placebos in the treatment of depression, and it would also point the way to understanding how to optimize treatment in clinical settings.

In 1976, Aaron Beck, a psychiatrist at the University of Pennsylvania, proposed a cognitive theory of emotions and emotional disorders – a theory that was to become the foundation for cognitive behavioural therapy for depression. According to Beck, fear is produced by the anticipation of harm, joy by the expectancy of positive events, and sadness by the sense that something important has for ever been lost. As a consequence, overcoming fear and depression requires changing the beliefs that have produced them.

The American President Franklin Delano Roosevelt once said that 'there is nothing to fear but fear itself'. It was a wise conclusion, especially from the standpoint of clinical psychology. Fear is indeed frightening. So much so that phobias – irrational fears of situations that are not dangerous – can be generated and maintained by the simple belief that one will experience intense fear. The panic and anxiety that are aroused in these disorders can be a simple, but intense, fear of fear.[56]

Just as fear is a frightening experience, depression is depressing. It is a terrible state of affairs, and many depressed people feel that they are trapped in it for ever. There may be other circumstances behind their depression, but depression about depression is certainly an important component.[57] Bringing their depression to an end requires instilling a sense of hope – a belief that their depression will not last for ever. For people who are depressed about depression, this change in expectation may be an essential component of effective treatment. If depression is a nocebo effect, then its treatment requires a positive placebo effect.

* * *

The evidence I have reviewed in this chapter indicates that placebos work for a wide variety of conditions. They can produce both positive and negative effects. They affect the body as well as the mind. They can be as strong as potent medications, and their effects can be lasting. We have also seen that placebos can produce negative effects. Furthermore, the nocebo effect may be an important factor in clinical depression – at least for some depressed people. For this reason, understanding the placebo effect is essential to understanding how to treat depression effectively. How do inert substances produce both therapeutic and detrimental effects? Chapter 6 provides an answer to this question.

6

How Placebos Work

If we are to harness the placebo effect and make use of it in clinical practice, we first have to understand how it works. A number of factors have been proposed as explanations of the placebo effect. These include the relationship between doctors and patients, the patient's beliefs and expectations, the production of opiates in the brain, and a phenomenon called classical conditioning, in which people come to associate pills and injections with therapeutic effects, just as Pavlov's dogs came to associate the sound of a bell with the presentation of food. In this chapter we look at how all of these processes combine to produce placebo effects, and we consider their implications for the treatment of depression.

You might find some of this material tough going, and if you are willing to take my word for the significance of these factors, you could just skip over these parts. But I thought it important to document my claims about how placebos work. Just as I have documented my claim that most of the antidepressant drug response is a placebo effect and that the remainder is in all likelihood an enhanced placebo effect, so here, too, I present the details of the research upon which my conclusions about the way placebos work are based.

THE THERAPEUTIC RELATIONSHIP

What happens when you go to your primary-care physician? Do you feel that he is engaged with you? Does she make frequent eye contact? Does he ask enough questions, and does she seem to listen when you answer them? Or does he seem impatient and rushed, spending more time looking at a laptop computer than at you? The way in which a clinician interacts with her patients can affect the outcome of treatment – and not just of treatment for mental-health problems, but treatment for physical conditions as well.[1]

Perhaps you have experienced the sense of well-being that a good 'therapeutic relationship' engenders. I know that I have. I had a doctor in New Jersey who had the most wonderful bedside manner. Dr Doubek – I called her Marnie – looked me in the eye when I spoke. She listened, she nodded, she showed concern. She did not seem the least bit hurried or rushed. And I do not know if she is aware of this, but at least once during each visit she touched me briefly on the arm while talking to me. I felt cared for, understood.

I trusted Marnie when I was her patient, and I still do. Two years after leaving New Jersey and moving to England, my wife and I wondered about the meaning of some medical test results we had obtained. We phoned Marnie to help us understand them, and even on the phone, with people who had not been her patients for more than two years, Marnie was forthcoming, patient and helpful. I only wish I could videotape the way in which she conducts her clinical sessions and have the DVD shown to all medical-school students.

I can't call Marnie's style of interacting with patients a placebo effect, because as far as I know none of the treatments she gave me were placebos. But it did make me feel better, and some of the research I describe in this chapter indicates that it can also promote wellness. For want of a better term, I will call this the 'Marnie effect'. The Marnie effect is the enhancement of treatment outcome that is produced by enhancing the therapeutic relationship.

The relationship between the medical practitioner and the

patient is, without any doubt, an important component of the placebo effect. Recently I was able to verify that hypothesis scientifically as part of a research team led by Ted Kaptchuk, an Associate Professor of Medicine at the Harvard University Medical School.[2] Kaptchuk is the most unusual associate professor at Harvard – or at any other university, for that matter. Not only does he not have a PhD or MD, but he does not even have a master's degree. Instead, after graduating from Columbia University with a bachelor's degree, he went to Macao, where he obtained an OMD – a Doctor of Oriental Medicine degree.

Kaptchuk returned from China a proponent of acupuncture and wrote *The Web That Has No Weaver*, the classic explanation of Chinese medicine for Western readers.[3] But over time he came to wonder whether the effects of acupuncture might be at least partly due to the placebo effect. To answer that question, he taught himself how to design research studies, and he did so well enough to obtain funding from the National Institutes of Health and publish more than 100 articles in leading medical journals. No wonder Harvard saw fit to hire and promote him, despite his rather unusual academic credentials.

Like real medicines, placebos show a dose-response relationship. The more you take, the greater the effect. Taking placebo pills four times a day provides greater relief from ulcers than taking only two a day,[4] and people who take their heart medication as prescribed live longer than those who do not, even if what they are taking is really a placebo that they have been given in a clinical trial.[5] Kaptchuk wondered whether the 'dose' of the therapeutic relationship could be altered just like the dose of a medication and, if so, whether this might affect the effectiveness of treatment. So he designed a study to find out and invited me to be one of the researchers.

We gave patients suffering from irritable bowel syndrome three 'doses' of a therapeutic relationship. The lowest dose was no relationship with the medical practitioner and no treatment at all. These patients were simply assessed and put on a waiting list, with the promise that they would receive treatment some weeks later. Another group of patients was given placebo acupuncture (using the fake needle that does not prick the skin, as I described in Chapter 5) with a 'low dose' of the therapeutic relationship. These patients were seen by a licensed acupuncturist who told them that because this was a scientific study, he had been instructed not to converse with them. A third group of patients received the 'high-dose' therapeutic relationship. Prior to starting the fake acupuncture treatment, the acupuncturist interviewed these patients for 45 minutes. He was warm and friendly with them, saying things like: 'I can understand how difficult your condition must be for you.' He took time to ponder the treatment plan and instilled a positive expectation by saying, 'I have had much positive experience treating irritable bowel syndrome and look forward to demonstrating that acupuncture is a valuable treatment in this trial.'

The results of this study showed that we were right about the effects of the therapeutic relationship. Six weeks after the beginning of treatment, patients given an enhanced therapeutic relationship reported significantly greater symptom reduction and better quality of life than those given the low-dose relationship, despite the fact that the difference in treatment was limited to the initial interview. Those in the wait-list group showed the least improvement of all.[6]

Our study showed that enhancing the therapeutic relationship boosts the placebo effect. Other studies have shown that the same thing happens when real treatments are delivered within the context of a caring relationship. When a doctor is warm, friendly, reassuring and confident in the effectiveness of the treatment, patients show greater symptom reduction and recover from illnesses more quickly.[7]

FEELING GOOD

How is it that the quality of the therapeutic relationship can enhance improvement, not only in a psychological condition like depression, but also in a physical disorder like irritable bowel syndrome? A clue to the answer to this question lies in one of my all-time favourite studies. You may have read E. M. Forster's book *A Room with a View* or seen the film that was based on it. As it turns out, having a room with a view not only makes a holiday more pleasant, but can also improve your health. Roger Ulrich, a researcher at the University of Delaware, divided patients who had just had gall-bladder surgery into two groups.[8] Patients in one group were given rooms with windows looking out over a park-like setting with trees and plants. The other patients were assigned rooms with views of a brick wall. Those given the rooms with the view of trees and plants required significantly less pain medication and were discharged from the hospital sooner than the others. Their nurses were more likely to describe them as doing well and being in good spirits. Patients in the rooms facing the brick wall needed more medication, took longer to be discharged and were described as upset and needing encouragement. In other words, feeling good psychologically makes you feel good physically. A warm and caring therapeutic relationship feels good. It leads the patient to feel hopeful rather than hopeless. It facilitates an expectation for improvement and that may, at least in part, explain its ability to facilitate healing.

The ability of emotions to affect health, for better and for worse, has been shown in other studies as well. Negative emotions, such as those induced by stress, can worsen physical health. They can increase blood pressure, impair the functioning of the immune system and increase the risk of death from heart disease.[9] There has been less research on the health benefits of positive emotions, but the research that has been done suggests that it can have curative effects. For example, people who are generally optimistic have lower blood pressure, better immune function and recover better from heart surgery. There is even some evidence that survival from cancer might be affected by emotional well-being.[10] So maybe Mr Wright's response to placebo treatment for cancer, in the case study I described in Chapter 5, was not merely a coincidence.

THE SPECIFICS OF 'NON-SPECIFIC' EFFECTS

Although the therapeutic relationship and positive emotions are clearly important, there are many instances of the placebo effect that they cannot explain. They cannot, for example, explain the effect of placebos in research settings in which students or other healthy volunteers have been asked to participate in return for money or course credit. In these studies there is no therapeutic relationship. Most importantly, emotions cannot explain the specificity of the placebo effect.

If you look in the medical literature, you will often see the term 'placebo' defined as a 'non-specific' treatment. What does it mean to say that a treatment is not specific? It could mean that the treatment is effective for many different disorders, rather than for only one particular condition. In this sense, placebos are indeed non-specific. Besides depression, placebos have been shown to affect anxiety, pain, ulcers, irritable bowel syndrome, Parkinson's disease, angina, autoimmune diseases, Alzheimer's disease, rheumatoid arthritis, asthma, gastric function, sexual dysfunction and skin conditions.[11] We know this from the thousands of studies in which placebos have been used as control conditions, against which the effects of medication have been evaluated, and from studies that were specifically designed to assess the placebo effect.

Although placebo effects are generally referred to as non-specific, there is also a sense in which they are very specific. The effect of the placebo is specific to the beliefs that people have about the substance they are ingesting. Placebo morphine, for example, reduces pain, whereas placebo antidepressants reduce depression. Even the side effects that people report when given a placebo tend to be the same side effects that are produced by the real drug.[12] In other words, the effect of a placebo is specific to the effect that the person expects it to have. When given placebo stimulants like decaffeinated coffee (presented as regular coffee), people feel more alert, and their heart rate and blood pressure may go up, but when given placebo tranquillizers, they feel more calm and relaxed, and their heart rate and blood pressure go down.[13] These opposite effects have been produced in studies with healthy volunteers as subjects. They were conducted in sterile laboratory settings, in which there was no therapeutic relationship at all. Furthermore, there is no reason to think that the healthy participants in these studies have any particular feelings about being given one or another type of placebo. The placebo effects found in these studies cannot be explained by the therapeutic relationship or by the positive emotional state it induces, but they can be explained by the subjects' beliefs and expectations about what they have been given. Some years ago, I coined the term 'response expectancy' to denote the expectations that are evoked by placebos, and this has since become an accepted factor in theories of the placebo effect.[14]

If placebo effects depended completely on the therapeutic relationship and patients' emotional states, it would not be possible for the same person to show placebo effects and nocebo effects at the same time. But they do. Sometimes the same person reports both therapeutic benefits and side effects from the same placebo.[15] Sometimes the more negative side effects they have experienced, the better they feel. That can happen because the side effects might convince them that they have been given a potent medication. Maybe their improvement was generated by their happiness over receiving what they believe to be an effective treatment for their condition, but this certainly would not explain their experience of side effects.

My former student Guy Montgomery, who is now a researcher at the Mount Sinai School of Medicine in New York City, demonstrated experimentally that placebo pain reduction cannot be completely explained by the patient's emotional state – or by any other factor that should affect a person's whole body instead of just part of it. He put a placebo cream on the index finger of one hand of his subjects and nothing at all on their other hand. Then he induced pain by putting heavy weights on the subjects' index fingers. He measured the placebo effect as the difference in the pain that the subjects felt in the finger on which he had put the placebo cream and the pain they felt in the untreated finger. Sometimes he put the weight on both fingers at the same time. At other times he tested each hand separately. That did not make a difference. He got the same placebo effect, regardless of whether he put the weight on both hands simultaneously or whether he did so sequentially.[16] Now, if placebo effects were produced by inducing positive emotions, or by reducing anxiety as has also been hypothesized,[17] then he should not have gotten a placebo effect when he put the weight on both hands at exactly the same time. Fingers do not feel anxious or happy; people do. So any effect on pain produced by their emotional state should have occurred in the fingers of both hands.

When Montgomery and I published our article, we thought we had disproven another theory of placebo effects – the theory that placebo effects are produced by the release of endorphins in the brain. In 1978 researchers at the University of California in San Francisco discovered that when placebos reduce pain, they may stimulate the release of endorphins.[18] Endorphins, the existence of which had only been discovered a few years earlier, are opioids that are produced naturally by the brain. Just like the opiates that are derived from opium – morphine and codeine, for example – endorphins reduce the sensation of pain. The University of California researchers reasoned that if placebos can mimic the effects of opiate drugs, maybe they do so by stimulating the release of the brain's endogenous opioids.

To test their hypothesis, the researchers gave placebo morphine intravenously to a group of patients who had just undergone dental surgery. An hour later they gave the patients a substance called naloxone. Naloxone is an opiate antagonist, which means that it blocks the pain-reducing effects of morphine and other opiates. In the California study, naloxone cancelled the pain-reducing effect of the placebo. This finding led the researchers to conclude that endorphins must have been involved in the production of the placebo effect in their post-surgical patients, a conclusion that has since been confirmed more directly by scanning people's brains during placebo treatment.[19]

Montgomery and I assumed that the release of endorphins in the brain would have general global effects throughout the entire body. It could not affect the perception of pain in just one part of the body, without affecting the rest of the body. Others shared our opinion, including Howard Fields at the University of California, one of the authors of the original naloxone study. But it turned out that we were wrong. Fabrizio Benedetti and his colleagues at the University of Turin Medical School repeated our study. Only this time he also assessed the effects of hidden infusions of naloxone, just as the University of California researchers had done. The results of Benedetti's study surprised everyone. He found a placebo effect despite having applied the pain stimulus to treated and untreated parts of the body simultaneously, just as Montgomery and I had found. That in itself was not surprising, but he also found that naloxone abolished this placebo effect completely. It seems that when expectancies of reduced pain lead the brain to release endorphins, these endogenous opiates can act on the specific part of the body towards which the expectancy is directed. I think it is safe to say that no one, with the possible exception of Benedetti and his collaborators, would have thought that possible.[20] Everyone else assumed that the pain-reducing effect of endorphins was global, affecting the person's entire body, rather than targeting specific locations.

Classical Conditioning

It has now been well established that expectancies play a central role in the production of placebo effects.[21] People's expectations of relief are not only correlated with how much benefit they report, but also with changes in the brain activity associated with the therapeutic benefit. These expectancies are formed and altered in many different ways. Our beliefs are influenced by parents, teachers, friends and colleagues, the advertisements we see on television and in newspapers and magazines, news programmes and documentaries, books and magazine articles. But the most effective way to alter beliefs and expectations is through direct experience.

The process by which experience affects our expectations is called classical or Pavlovian conditioning. I assume you already know about Pavlovian conditioning, but a brief review may nevertheless be useful. Classical conditioning was discovered at the turn of the 20th century by the Russian scientist Ivan Pavlov.[22] Pavlov was a distinguished physiologist who had been awarded a Nobel Prize in 1904, not for his work on conditioning, which he was just beginning and which few people knew of in 1904, but for his research on the physiology of digestion in dogs.

In 1897 one of Pavlov's doctoral students discovered that after stimulating dogs to salivate by having them smell a glass of carbon bisulphide, the dogs began to salivate when presented with a glass of plain water. Eventually this discovery changed the direction of Pavlov's research. He began using many different stimuli to induce salivation, including tuning forks, musical scales, tapping on a glass and, most famously (although much later), ringing a bell. These stimuli were paired with food in Pavlov's studies. The food was called an unconditioned (or unconditional) stimulus, because it evoked salivation as a reflex, even if there had been no conditioning at all. The bell, tuning fork or musical scale was termed a conditioned (or conditional) stimulus, because it provoked salivation only after it had been associated with food.

The 1897 study in which Pavlov's student substituted a glass of water for the carbon bisulphide that had been used to stimulate the dogs to salivate shows the relevance of classical conditioning to the placebo effect. The glass of water was a placebo. Although it was inert, it looked exactly like the substance that had led the dogs to salivate.

Here then is the classical conditioning account of the placebo effect. People experience getting better after having been given active medications. These medications are always administered in some kind of vehicle – in a pill, a capsule or by injection. Eventually the pills, capsules and injections become associated with the effects of the drug and are able to reproduce those effects as 'conditioned responses'.

Some years ago a team of Australian researchers conducted an ingenious series of studies showing how Pavlovian conditioning could strengthen the placebo effect.[23] They told the subjects in their study that they were testing a powerful fast-acting analgesic cream, which was actually a placebo. Then they repeatedly stimulated the subject's arms with a pain generator that drives positive potassium ions into the skin, causing a painful cramping sensation. Sometimes the arm had been treated with the placebo cream; sometimes it had not. To strengthen the placebo effect, the experimenters surreptitiously lowered the intensity of the pain stimulus whenever the placebo cream had been applied. Then they tested the effect of this conditioning procedure by turning the intensity of the pain generator back up to its original level, so that exactly the same level of stimulation would be used, regardless of whether or not the placebo had been applied. What they found was that this conditioning procedure increased the placebo effect substantially. The subjects who had experienced reduced stimulation, but without knowing that the intensity had been reduced, later reported significantly more placebo pain reduction than a control group that had not undergone this conditioning procedure – this despite the fact that the intensity of the pain generator had been turned back up to full level when the effect of the conditioning procedure was tested.

In a later study, my colleagues and I showed that we could gain an exquisite degree of control over experienced levels of pain by using this conditioning procedure.[24] We dabbed a liquid mixture from a medicinal bottle bearing the label 'Trivaricaine-A' on to one area of each person's arm, and we applied liquid from a bottle labelled 'Trivaricaine-B' to a different part of the arm. We told people that these bottles contained different strengths of the same topical anaesthetic and that we were testing their efficacy. Of course, both liquids were placebos. (I had been tempted to label our placebo 'Prevaricaine', but my colleagues talked me out of it.) After the placebos had been applied and given time to take effect, we administered pain stimulation to each of the two areas on the arm. We also stimulated a third area of the arm, on which we had applied plain water.

During our conditioning procedure, without the subject knowing it, we manipulated the intensity of the pain stimulus that we applied to each area. We administered an intensely painful stimulus to the area where we had put plain water, slightly less intense stimulation to the Trivaricaine-B area and much lower intensity to the Trivaricaine-A area. Then, to test the effects of this conditioning procedure, we applied the same level of pain to all three areas. As we had expected, subjects reported the greatest amount of pain in the control area where we had applied plain water, less pain in the Trivaricaine-B area and even less in the Trivaricaine-A area.

You might suspect that the subjects in these conditioning studies might have been lying to the experimenters – just telling us what we wanted to hear. But we now know that this is not the case. Tor Wager used this conditioning procedure before he scanned the brains of subjects while they were given painful stimulation with and without the benefit of the placebo to which they had been conditioned.[25] The subjects not only reported experiencing less pain, but they also showed reduced activity in the pain network of the brain.

Expectancy and Conditioning

For many years there was a debate in the scientific literature about whether placebo effects are produced by conditioning or by expectancy. Now the answer seems as clear as Pavlov's bell. Both factors are involved. Specifically, conditioning is one of the means by which expectancies are produced and altered.[26] After repeatedly being given food just after hearing a bell ring, the dog comes to expect to be fed whenever it hears the bell. After successful treatments with active drugs, we come to expect drugs to have positive effects. We can form expectations like this even without direct experiences of this sort – for example, by being told of the effectiveness of a medication – but direct experience (that is, conditioning) is the most convincing source of information.[27]

Guy Montgomery and I confirmed the role of expectancy in conditioned placebo effects in a study that we conducted while he was working on his doctoral dissertation under my supervision at the University of Connecticut.[28] We repeated the conditioning procedures that the Australian researchers had developed, but we added a new twist. In addition to having subjects who did not know that we were turning down the level of the pain stimulus during the conditioning trials, we also ran a group in which we told people that we were turning down the stimulus intensity. Our idea was that this group of people would have the same classical conditioning experience, in which pain reduction would be paired with the placebo tincture, but they would know that the reduced pain they were feeling was not due to the placebo. We reasoned that if conditioning were an automatic process that does not depend on people's expectancies, then even this 'informed' group should show the conditioning effect by experiencing a greater placebo effect when we turned the pain generator back up to full intensity. On the other hand, if the effect of conditioning occurs because people come to expect less pain when given the placebo, then the 'informed' group should not benefit from the conditioning procedure.

The results of our study clearly established that conditioning enhanced the placebo effect by changing people's expectations. As had the Australian researchers, we found that conditioning increased the placebo effect for the subjects that we had kept in the dark about our manipulation. They came to expect less pain, and they subsequently experienced less pain. But conditioning had no effect at all on the subjects who were told that we were lowering the intensity of the pain stimulus. Knowing that we had lowered the stimulus intensity, they did not come to expect less pain where the placebo had been applied, and since they did not expect less pain, they did not experience it when the intensity was turned up again.

Researchers at the University of Manchester recently replicated our study and used an EEG to record their subjects' brain activity.[29] Similar to our results, they found that lowering the intensity of the pain stimulus enhanced the subsequent placebo effect on self-reported pain, but only if the subjects did not know that the stimulus intensity had been reduced during the conditioning session. They also showed that this effect was not just because the subjects were telling the experimenters what they wanted to hear. Instead, the reports of reduced pain were accompanied by reductions in brain activity.

I am tempted to conclude that the only direct effect of conditioning is to change expectancies, and that it never has automatic effects that aren't based on what the person believes. But that would be going too far. Classical conditioning can be seen in organisms as simple as the California sea slug, and I would be very reluctant to attribute thought processes to anything with such a simple nervous system. People have expectations, and I am convinced that dogs do as well, but I draw the line at slugs. Consciousness most certainly evolved from simpler unconscious processes, and Pavlovian conditioning is one of those processes. In lower organisms the effect of conditioning on behaviour is direct. In higher organisms, conditioning provides information that can be used to decide on a course of behaviour.

Still, there seem to be some vestiges of automatic conditioning effects that affect the placebo response and are not based on expectancy. Consistent with the result of the study I did with Guy Montgomery, Fabrizio Benedetti and his colleagues have shown that conditioned placebo effects on conscious experiences like pain depend on people's expectations, but they also found a conditioning effect on hormonal secretion that could not be blocked by preventing a change in expectancy.[30] Automatic conditioning effects like these are the exception rather than the rule. They seem to be limited to unconscious processes like hormone secretion. The effects of conditioning on conscious processes like pain depend on people's expectations.

HARNESSING THE PLACEBO EFFECT IN CLINICAL PRACTICE

During the 1980s the National Institute for Mental Health (NIMH) in the United States sponsored a massive, multi-centred research programme to evaluate the effectiveness of antidepressants and psychotherapy in the treatment of depression.[31] Before beginning treatment, each patient was asked the following question: 'What is likely to happen as a result of your treatment?' They were asked to respond to this question on a five-point scale, on which low expectancy for improvement was represented by the sentence 'I don't expect to feel any different' and high expectancy was represented by 'I expect to feel completely better'. Patients' answers to this question predicted their therapeutic outcome. Those who expected to feel better improved the most, and those who did not expect to feel better got the least benefit from treatment. Furthermore, the effect of expectancy on treatment outcome was independent of which treatment they had been given. Regardless of whether they had been treated with antidepressant medication, psychotherapy or a placebo, patients who expected to get better showed the most improvement.[32]

One lesson to learn from the findings of the NIMH Treatment of Depression Collaborative Research Program is that people who are depressed need to be convinced that the treatment they are being offered – whatever it is – is effective and that it offers them hope for what they may until then have considered a hopeless situation.[33] Some people come into treatment with positive expectations, but others do not, and unless the clinician puts effort into changing negative expectations at the outset, treatment is not likely to be very effective. This is yet another reason for concluding that the effects of medication on depression are placebo effects. If the effect of these drugs were not at least partly a placebo effect, they would work – like antibiotics and hypoglycaemics – regardless of patients' expectations.

As I discussed in Chapter 5, depression is partly a nocebo effect, in the sense that it can be produced by negative expectations about oneself and the world.[34] The way in which these negative expectations develop and produce their negative effects provides some clues as to how they can be reversed. Expectancy effects grow, feeding upon themselves. One reason this happens is that our subjective states – our feelings, moods and sensations – are in constant flux, changing from day to day and from moment to moment. The effects of these fluctuations depend on how we interpret them, and our interpretations depend on our beliefs and expectations. When we expect to feel worse, we tend to notice random small negative changes and interpret them as evidence that we are in fact getting worse. This interpretation makes us actually feel worse, and it strengthens the belief that we are getting worse, leading to a vicious cycle in which our expectations and negative emotions feed on each other, cascading into a full-blown depressive episode. This is the process by which relapse can occur when someone is taken off antidepressants. Positive expectancies have the opposite effect. They can set in motion a benign cycle, in which random fluctuations in mood and well-being are interpreted as evidence of treatment effectiveness, thereby instilling a further sense of hope and countering the feelings of hopelessness that are so central to clinical depression.

In 1998 a group of researchers at Columbia University provided evidence for this snowball effect.[35] Recall that at the beginning of most clinical trials, all patients are given placebos for a week or two in what is called a placebo run-in period. The Columbia researchers looked at what happened during this run-in period and found that the patients who improved during it were the ones most likely to improve later in the trial, regardless of whether they were then given antidepressants or placebos. In a subsequent study, researchers at the University of California at Los Angeles (UCLA) confirmed these results by identifying changes in regional brain activity during the placebo run-in period that predicted improvement when patients were given medication.[36]

Positive expectations can also backfire, and clinicians need to be careful about the beliefs they foster. The key to preventing this is to understand that expectancies of improvement have different facets and that these different kinds of expectancy can function independently.[37] One of these facets is the amount of change the person expects. I might expect a complete cure, or I might expect no change at all. This is the kind of expectancy that was correlated with improvement in the NIMH Treatment of Depression Collaborative Research Program. A second aspect of a person's expectancy is the confidence with which it is held. I may be absolutely certain that I will change, or I might be very unsure. This is the kind of expectancy we alter when we tell people in a clinical trial that we might give them a placebo. As we have seen, lowering expectancy in this way decreases the effect of treatment.[38]

But there is also a third aspect to our expectancies for improvement, and that is the speed with which the change is expected. I might expect change to occur almost immediately, or I might expect it to happen gradually over time. This is the aspect of our expectations that can have paradoxical effects. When people have unrealistic expectations, when they expect too much change to occur too quickly, their expectations are likely to be shattered.

There is yet one other aspect of expectancy to which we need to attend. Patients might expect change to occur automatically, without them having to do anything to bring it about. This is the expectation fostered by drug treatment. One does not expect to have to do anything but take the drug for it to have its effect. Alternatively, one might expect to have to work actively to bring change about, rather than to wait passively for it to occur. Data from the NIMH study showed that when patients expected treatment to work, they also got more involved in working with the clinician to help bring those changes about, and the more actively involved they became with the treatment process, the more they improved.

Here then are the kinds of expectations that are most likely to lead to therapeutic improvement and that should be fostered by clinicians. To maximize therapeutic outcome, it is best to be confident in the effectiveness of treatment, to expect substantial change, but also to expect that change to occur gradually. The changes are likely to be subtle at first, and to increase over time. It is also helpful to understand that change is not automatic; one has to work to bring it about. These are the kinds of expectations that are best suited to interrupting vicious cycles and replacing them with benign ones.

As we have seen, expectancy is not the only factor that encourages placebo effects. There is also the effect of the therapeutic relationship – what I have called the Marnie effect. A caring therapeutic relationship enhances the patient's confidence, and in so doing also fosters positive expectations. But it can also affect patients' well-being in ways that are independent of expectancy. A positive therapeutic relationship feels good, and feeling good counters depression and may also have more general health benefits.

Is the Marnie effect too much to ask for in primary care? Doctors are busy. They have large caseloads. The amount of time they have available for each patient is limited, and making more time available would cost money. Still, this might be money well spent, given the potential benefits for health and well-being that it could produce. In the long run it might even be cost-effective. It has the potential to reduce the number of primary-care visits that patients need and the number of referrals that need to be made.

* * *

Enhancing expectations and strengthening the therapeutic relationship might enhance the outcome of treatment. But what treatment? Antidepressants may be nothing more than active placebos, producing side effects through chemical means and therapeutic effects only through psychological means. In the next chapter we consider the various options that are available for the treatment of depression. As we shall see, some of these treatments go well beyond the Marnie effect in the treatment of clinical depression.

7

Beyond Antidepressants

We are faced with a dilemma. Millions of people suffer from depression. Many of them get better when treated with antidepressants, whereas left untreated, they do not show much improvement at all.[1] The problem is that antidepressants have turned out to be not much more effective than placebos.

The placebo effect in the treatment of depression is very large, and it is likely to be even larger in clinical practice than it is in clinical trials. In clinical trials, people are told that they might be getting a placebo, and this knowledge diminishes the placebo effect.[2] In clinical practice, on the other hand, people know they are getting an active medication and, trusting their doctors, they are more confident that they will improve.

So what are we to do? Perhaps we should continue prescribing antidepressants, even if they are placebos, given that they are very effective placebos. As one psychiatrist put it, 'It matters little whether the patient responds because of a placebo effect or the specific pharmacological actions of the drug, as long as he / she gets better'.[3] But there is a problem with this solution. Antidepressants may be placebos, but unlike the placebos that are used in almost all clinical trials, they are not inert. Instead they are active drugs, and as such they produce effects that are not placebo effects. The problem is that many of these real drug effects are harmful side effects rather than beneficial therapeutic effects.

The side effects of antidepressants are a serious problem. Many depressed patients find them so intolerable that they stop taking their medication. This leads many of them to drop out of clinical trials within a few weeks after beginning treatment, which is why most of these trials are short, lasting only between four and eight weeks. Drug companies have directed most of their efforts not towards finding more effective antidepressants – differences in effectiveness between one antidepressant and another are clinically insignificant – but towards developing drugs that have fewer side effects and will therefore be more tolerable. This is the advantage of SSRIs and other 'new-generation' antidepressants over older drugs that were used to treat depression. The new drugs are not more effective, but they do have fewer side effects.

Although SSRIs have fewer side effects than older antidepressants, the list of adverse events is still substantial – substantial enough to preclude their use as 'active placebos'. Among the reported side effects of SSRIs, Eli Lilly (the manufacturer of Prozac) lists the following in their official Summary of Product Characteristics: sexual dysfunction, headaches, nausea, vomiting, insomnia, drowsiness, diarrhoea, sweating, dry mouth, seizures, mania, anxiety, impaired concentration, panic attacks, fatigue, twitching, tremors, dizziness, anorexia, dyspepsia, difficulty swallowing, chills, hallucinations, confusion, agitation, photosensitivity, urinary retention, frequent urination, blurred vision, hair loss, pain in the joints, hypoglycaemia, rashes and serious systemic events involving the skin, kidneys, liver or lungs. Furthermore, these are only the more common side effects that have been associated with SSRIs. Lilly reports other undesirable effects, such as hepatitis and haemorrhages, as occurring 'rarely'.

Each of these side effects is experienced by only a minority of patients taking the drugs. About 15 per cent of patients taking SSRIs report headaches, and the same number complain of nausea. Diarrhoea, dizziness and insomnia are reported by 10 per cent of patients on SSRIs. But while only a minority report any particular side effect, the number of patients who report suffering from at least one of them is quite high, ranging from half of the patients to the vast majority of them, depending on how these adverse events are assessed.[4]

The risk of potentially serious side effects should be enough to preclude the prescription of antidepressants for their placebo benefit, but this is not the only hazard associated with these medications. On 19 July 2006 the FDA issued a public-health advisory warning that, when taken in conjunction with other drugs that can affect serotonin levels, antidepressants can induce a life-threatening disorder called the 'serotonin syndrome'.[5] The serotonin syndrome is caused by an excess of serotonin in a person's body.

One way of inducing the serotonin syndrome is to take more than one antidepressant at the same time, but it has also been associated with the concurrent use of other drugs, including over-the-counter headache remedies and cough suppressants. Other drugs that have been implicated in producing this potentially fatal condition when taken together with antidepressants include analgesics, antibiotics, herbal medications, appetite suppressants and street drugs like ecstasy, cocaine and LSD.[6] Symptoms of serotonin syndrome include restlessness, hallucinations, loss of coordination, a racing heart, rapid changes in blood pressure, fever, nausea, vomiting and diarrhoea.

Suicidal thoughts are one of the symptoms of depression. Paradoxically, one of the best-publicized dangers of SSRIs is their potential to increase the risk of suicide. This heightened risk is especially well established for children, adolescents and young adults. In their most recent analysis of the data, the FDA concluded that, when compared to placebos, SSRIs double the risk of suicidal thoughts and behaviour in depressed patients up to the age of 24.[7] There also seems to be an increased risk for people who are older than 24, but the interpretation of these data is still disputed.[8]

Not only has the connection between SSRIs and suicide been well established, but we also have some idea how SSRIs might produce this increased risk. The American psychiatrist Peter

Breggin has documented how SSRIs can provoke an agitated, restless state called akathisia, which some people describe as feeling like jumping out of their skin.[9] It is often in this state that people on SSRIs become violent and aggressive towards themselves or others.

I first learned of the akathisia connection on 2 February 2004, at the FDA hearing that resulted in the addition of the 'black box' warning to SSRI labelling information. Along with my colleague David Antonuccio, I had been invited to testify at the hearing about the efficacy – or lack of efficacy – of SSRIs as a treatment for childhood depression. It was there that I first heard the heart-wrenching stories of parents whose children had committed suicide, of a 12-year-old boy who had murdered his grandparents with a shotgun, and of a woman who had shot her jaw off while taking SSRIs. It was also at that hearing that I first learned of the clinical studies in which akathisia was turned on and off by Prozac. In one of these studies, three patients, aged 25–47 years, who had attempted suicide while on Prozac and then been taken off of the drug, were given Prozac again to see what would happen. All three of them developed severe akathisia and reported feeling suicidal again. The manic feelings subsided, as did their suicidal thoughts, when the drug was discontinued again.[10]

WARNING: DO NOT DISCONTINUE ANTI-DEPRESSANTS WITHOUT CONSULTATION

Understandably, learning that the benefits of antidepressants are largely due to the placebo effect, some depressed patients may be tempted to stop taking their medication. With this in mind, I have asked the publishers to highlight the following warning in bold typeface. It is akin to the black-box warning about the increased risk of suicide that is contained in the approved labelling for antidepressants: **Antidepressant medication should not be discontinued without first discussing it with your doctor**.

The reason for this warning is that abrupt cessation of SSRIs produces withdrawal symptoms in about 20 per cent of patients. Symptoms of withdrawal from antidepressant medication include gastrointestinal disturbances (abdominal cramping and pain, diarrhoea, nausea and vomiting), flu-like symptoms, headaches, sleep disturbances, dizziness, blurred vision, numbness, electric-shock sensations, twitches and tremors. Abrupt withdrawal can also produce symptoms of depression and anxiety, which can occur within hours of the first missed dose of the drug.[11] Withdrawal symptoms are sometimes mistaken for a relapse, leading patients to resume antidepressant medication and to conclude that they need it in order to remain free of depression. Technically, this is not considered 'addiction', but it does seem awfully close.

If you are currently taking an antidepressant drug, if you are happy with its effects, and if side effects are not causing undue problems, you might be best advised to continue taking it. As the saying goes, 'If it ain't broke, don't fix it.' On the other hand, if the drug is not producing sufficient benefit or if you are troubled by the side effects, you might consider alternative approaches to managing depression. I discuss some of these alternatives later in this chapter.

If you do decide to discontinue drug treatment, talk to your doctor first. It will be important to taper your medication gradually, rather than stopping abruptly. The book *Coming Off Antidepressants* by Joseph Glenmullen of the Harvard Medical School is an excellent source of information on how to discontinue antidepressant drug treatment.

PRESCRIBING PLACEBOS

If the placebo effect in depression is so powerful, perhaps we should just prescribe inert placebos to depressed patients. They have been tested in thousands of clinical trials, they are the standard against which all other medications are evaluated, and they are safe enough to be taken by pregnant women, small children, the infirm and the elderly. You might think I am merely being facetious in suggesting this, but it has been recommended seriously and appears to be practised frequently. The *British Medical Journal* has published surveys in which doctors in the US and Israel were asked whether they sometimes prescribed placebos intentionally.[12] Approximately half of them responded that they did, mostly in the form of over-the-counter analgesics and vitamins.

The obvious problem with prescribing placebos is the fact that it generally entails deception. When physicians prescribe placebos, they don't tell their patients that the treatment is a placebo.[13] Instead, the patients are led to believe that they are receiving an active treatment. This raises a serious ethical question. Is it ethical to deceive patients if the deception is likely to make them better?

Two NHS physicians, Rudiger Pittrof and Ian Rubenstein, have argued that the use of placebos can be ethically justifiable and that it can be done without – strictly speaking – deceiving patients. The gist of their argument is that placebos work for some conditions (notably depression) and that this makes it possible to remain 'within the spirit of scientific, evidence-based medicine' when prescribing them. In fact, they suggest that it might be unethical to withhold placebo treatment that has been shown to be effective. Even in conditions for which placebos are not as effective as active medications – as in the treatment of sexual dysfunction in men, for example – the side effects and dangers of drug interactions could be avoided by prescribing placebos, and this might make placebo treatment preferable to many patients. Pittrof and Rubenstein recommend giving patients a choice between a possibly more effective treatment that has a greater likelihood of side effects and a somewhat less effective treatment (placebo) that has fewer side effects – without, of course, telling them that it is a placebo.

As sympathetic as I am to Pittrof and Rubenstein's arguments, I remain unconvinced. It may be possible to avoid technically lying to patients when administering placebos, but that just makes the deception implicit rather than explicit. Patients are still led to believe that they are getting a pharmacologically active treatment, when in fact they are not.

Perhaps it is my background as a psychotherapist that leads me to be concerned about the widespread practice of deceptively giving patients placebos. As a therapist, I learned that one of the principal factors in the success of treatment is the relationship between the doctor and the patient. Trust is one of the central components of the therapeutic relationship, but trust has to be earned. When it is betrayed, it is lost. So my concern is as much practical as it is ethical. When doctors deceive their patients, they violate their patients' trust. In the long run they will lose it and, in so doing, they will lose one of the most effective weapons in their clinical arsenal.

PLACEBOS WITHOUT DECEPTION

When given in the guise of active medications, placebos can produce powerful effects, but how potent would they be if the patients knew they were taking placebos? Is it possible to produce a placebo effect without deception?

In 1965, Lee Park and Lino Covi, two young psychiatrists at the Johns Hopkins University hospital, undertook a study that was aimed at answering this question.[14] Their surprising conclusion was that placebos can be given openly, without deception, and still be effective. Park and Covi gave placebo pills to 15 psychiatric outpatients and told them that the pills were placebos. More specifically, they told the patients: 'Many people with your kind of condition have . . . been helped by what are sometimes called "sugar pills", and we feel that a so-called sugar pill may help you, too. Do you know what a sugar pill is? A sugar pill is a pill with no medicine in it at all. I think this pill will help you as it has helped so many others. Are you willing to try this pill?'

With instructions like these, one might expect patients to become angry or insulted, to refuse to take the pills or at least to feel sceptical, even if reluctant to express their scepticism.

We would certainly not expect them to improve. Even the researchers who conducted the study did not expect to find much of an effect. 'It was to just be a very small pilot trial to learn if patients would actually go along with us and to see if any subjects actually benefited,' Lee Park recalls.

The researchers were amazed by the results. All but one of the 15 patients agreed to take the placebo pills. But did they actually take them? To find out, Park and Covi counted how many pills were left at the end of the first week. The pill count indicated that all 14 who had agreed to take the placebo pills had in fact taken them as prescribed. More impressive, all of them reported feeling better at the end of the study.

How did the placebo produce improvement in these patients? To find out, Park and Covi asked them what they thought about the pills they had been given. Eight of the patients suspected that the clinicians had lied to them and that the pills contained active medication. Three of these patients reported side effects that may have encouraged their suspicion. One patient concluded that the pills could not have been placebos because they worked better than medications that she had taken previously. Other patients were sure that the pills were in fact placebos, just as their doctors had told them they were. These patients also got better. In fact, one patient who was afraid of getting addicted to active medication expressed relief at having been given a placebo and asked to be allowed to continue taking her sugar pills after the experiment was over.

The Park and Covi study is certainly tantalizing, and it is a shame that no one has ever followed up on it with further research, because it is also a flawed study, and it is difficult to draw conclusions from it. The biggest problem is that there was no control group. The patients might have improved just as much even if they had not been given the placebo pill. Still, the fact that most of the patients complied with the non-deceptive placebo treatment instructions, and that some later attributed their improvement to having taken a 'sugar pill', suggests that the use of placebos need not be deceptive in order to be effective.

In 1965, when Park and Covi's study was published, placebos were just becoming a standard feature of medical research, and the general public was not as aware of them as they are now. I may be wrong, but I suspect that patients today would not be as amenable to the idea of taking a 'sugar pill' as they seem to have been then. Still, there are rationales for knowingly taking placebos that might be effective today. As I described in Chapter 6, classical conditioning is one of the factors behind the placebo effect. Classical conditioning is the phenomenon in which a neutral stimulus (such as a bell, buzzer or placebo pill) comes to evoke a reaction that had been produced by something else (food or active medication) with which it has been associated. Most of these conditioned responses are due to the beliefs and expectations that are produced by the conditioning process, but some of them are also automatic.[15] They can occur even without the person's conscious awareness. So perhaps taking a placebo pill is a smart idea after all, even if you know it is a placebo. The pill can function as a conditioned stimulus – as it is called in the scientific literature – triggering a therapeutic reaction because of your previous positive experience with active medications.

It may indeed be possible to give people placebos openly, without deceiving them, and still obtain good results. But there is a good argument for not doing this: we don't have to. There are alternatives to the prescription of either antidepressant drugs or placebos. These alternative treatments mobilize the placebo effect, and some of them may do much more than this, but they carry neither the side-effect risks of active drugs nor the ethical risks of deception. I explore these alternatives in the rest of this chapter.

PSYCHOTHERAPY: THE QUINTESSENTIAL PLACEBO

Of all the alternatives to antidepressant medication, psychotherapy is the most thoroughly researched. It has been the subject of hundreds of studies, which have been summarized in scores of meta-analyses. Indeed, there have been so many meta-analyses of the psychotherapy outcome research that there are even systematic reviews of the meta-analyses – that is, reviews of the reviews.[16] The results of these clinical trials, meta-analyses and reviews point to one inescapable conclusion. Psychotherapy works for the treatment of depression, and the benefits are substantial. In head-to-head comparisons, in which the short-term effects of psychotherapy and antidepressants are pitted against each other, psychotherapy works as well as medication. This is true regardless of how depressed the person is to begin with. It works for people who are moderately depressed, those who are severely depressed and even for patients who are very severely depressed.

Psychotherapy looks even better when its long-term effectiveness is assessed.[17] Formerly depressed patients are far more likely to relapse and become depressed again after treatment with antidepressants than they are after psychotherapy. As a result, psychotherapy is significantly more effective than medication when measured some time after treatment has ended, and the more time that has passed since the end of treatment, the larger the difference between drugs and psychotherapy. This long-term advantage of psychotherapy over medication is independent of the severity of the depression. Psychotherapy outperforms antidepressants for severely depressed patients as much as it does for those who are mildly or moderately depressed.[18]

There are a number of different psychotherapies for depression. The most common and thoroughly researched of these is cognitive behavioural therapy, known as CBT for short. As implied by its name, CBT has two components, a behavioural component and a cognitive component. The behavioural component of CBT is emphasized during the early stages of treatment, especially for severely depressed patients. It focuses on planning daily life activities, with special attention to activities that have the potential to provide pleasure and a sense of mastery and accomplishment. The cognitive component of CBT is based on the premise that emotions are not caused by the things that happen in our lives, but rather by the way in which we interpret those events. In other words, it focuses on the meanings that events have for us, and it is supposed to work by changing those meanings. It involves examining and challenging the negative thoughts that may promote and maintain depressed feelings – thoughts like 'I am a failure', 'I'm stupid' or 'no one will ever love me'. Depressed patients are asked to monitor the thoughts that spontaneously pop into their minds, and then, together with their therapists, they examine these conclusions, evaluate them logically and test them. The therapist and the patient work together as if they were research collaborators. They treat the patient's negative thoughts as hypotheses that can be tested and revised in the light of evidence and reason.

In the past, the cognitive and behavioural components of CBT were referred to different types of therapy – cognitive therapy and behaviour therapy. But it soon became clear that there were few (if any) differences between them. When treating depression, behaviour therapists worked on producing cognitive as well as behavioural change, and cognitive therapists used behavioural as well as cognitive techniques.[19] The distinction between behaviour therapy and cognitive therapy for depression was based on differences in theory rather than practice, and it has now pretty much disappeared.

Although it has received the most attention, CBT is not the only form of psychotherapy that is effective for depression. Other psychological treatments include interpersonal psychotherapy, short-term psychodynamic therapy and non-directive supportive therapy. Interpersonal psychotherapy focuses on problems that arise in interpersonal relationships, such as marital conflict, the loss of a loved one and social isolation.[20] Short-term psychodynamic therapy focuses on acquiring insight and understanding of unresolved conflicts arising from the person's childhood. It is based on Freud's psychoanalytic theory, but requires only months, rather than the years it takes for a full psychoanalysis.[21] Non-directive supportive therapy provides a warm, supportive atmosphere in which the depressed person can explore life issues in the presence of a caring professional. It is based on the premise that people have within themselves the ability to work through their psychological issues and to grow towards fulfilment and well-being. All that they need is a caring context in which they can feel safe enough to explore their inner world.[22]

Researchers comparing the effectiveness of these various psychotherapies have found some significant differences.[23] In general, cognitive behavioural treatments seem to be more effective than psychodynamic therapy, and non-directive supportive therapy seems less effective than any of the others. For very severely depressed people, interpersonal psychotherapy and the behavioural components of CBT seem particularly effective in the short run at least, and the long-term effects of CBT are particularly impressive, especially when compared to the long-term effects of antidepressants. For the most part, the differences in effectiveness of these therapies are not very large, and people who are depressed might well make a choice about which to seek on the basis of how much sense the treatment makes to them.

Psychotherapy, Medication or Both?

Psychotherapy has a number of advantages over medication. The most obvious is that it is not a drug, which means that it does not have the side effects or other risks that are associated with taking drugs. A second advantage is that it can be used safely to treat depression in children, adolescents and young adults, for whom antidepressants increase the risk of suicide.[24] A third advantage is that people are less likely to drop out of psychotherapy prematurely than they are to stop taking antidepressants, and this seems to be particularly true for patients with severe to very severe levels of depression.[25]

The greatest advantage of psychotherapy over medication is that it reduces the likelihood of relapse after having got better. In 2005 a group of Dutch researchers conducted a clinical trial in which they examined the effect of adding cognitive therapy to 'treatment as usual' in a group of patients with a history of recurrent depression.[26] These are the people who are most likely to relapse after treatment, because they are the ones who have relapsed in the past. The researchers found that among patients who had suffered five or more prior bouts of depression, cognitive therapy reduced the rate of relapse from 72 per cent to 46 per cent over a two-year period, and this benefit was independent of whether the patients took medication during the follow-up period.

The most impressive demonstration of the long-term benefits of psychotherapy comes from a study conducted by a group of Italian researchers led by Giovanni Fava at the University of Bologna.[27] Over a six-year period, they followed patients who had been successfully treated with antidepressants and then gradually taken off them. Half of the patients were given ten half-hour sessions of cognitive behaviour therapy (CBT). The others were also seen by the psychiatrist for ten half-hour sessions, but they were not given the actual therapy during these sessions. Instead, they received 'clinical management'. During these clinical-management sessions, the psychiatrist reviewed the patients' current state, discussed any problems that had arisen since the previous session and provided an opportunity for patients to express their feelings. In other words, patients in the control group were provided with all of the 'non-specific' placebo characteristics of psychotherapy, without any of the components that are specific to CBT. There was no attempt to help these patients schedule the activities of everyday life or to examine or change their negative beliefs and expectations. The results of this trial were dramatic. Six years after the ten-session treatment, 60 per cent of the patients who had been give CBT were symptom-free, compared to only 10 per cent of those who had only received clinical management.

Why does psychotherapy – either alone or in combination with antidepressants – have more lasting effects than medication? If you take antidepressants and get better, you are likely to attribute your improvement to the medication. So when you stop taking it, you might expect to get worse again. In Chapter

5 we saw that placebos can have negative as well as positive effects, in which case their effects are called nocebo effects, rather than placebo effects. Getting off medication may trigger relapse as a nocebo effect. This might not happen all at once. Instead, the normal slumps in mood that come with the stresses and strains of life might be interpreted as indications that depression is returning, beginning the vicious cycle that I described in Chapter 6, in which expectations and emotional slumps feed on each other, leading eventually to a relapse.

When people recover from depression via psychotherapy, their attributions about their recovery are likely to be different than those of people who have been treated with medication. Psychotherapy is a learning experience. Improvement is not produced by an external substance, but by changes within the person. It is like learning to read, write or ride a bicycle. Once you have learned, the skill stays with you. People do not become illiterate after they graduate from school, and if they get rusty at riding a bicycle, the skill can be reacquired with relatively little practice. Furthermore, part of what a person might learn in therapy is to expect downturns in mood and to interpret them as a normal part of life, rather than as an indication of an underlying disorder. This understanding, along with the skills that the person has learned for coping with negative moods and situations, can help to prevent a depressive relapse.

If both drugs and psychotherapy alleviate depression, maybe the combination of the two would work even better. This could be true even if the effects of antidepressants are placebo effects. As we saw in Chapter 4, taking two placebos can be more effective than taking only one.

There does, in fact, seem to be an advantage in combining antidepressants with psychotherapy, even in the short run, but the extra benefit of combining both treatments seems to be relatively small, and there is a catch. The advantage of combining treatments depends on whether you compare the combination treatment to drugs alone or psychotherapy alone. Combining psychotherapy and medication is better than just taking antidepressants, but it is not better than psychotherapy without drugs.[28] In other words, if you are in psychotherapy, there is no advantage to be gained by also taking antidepressants. On the other hand, if you are treated with antidepressants, you will be better off if you also get psychotherapeutic treatment. But since the effect of psychotherapy alone is as great as the combined effect of psychotherapy and antidepressants, why bother with the drugs?

Is Psychotherapy a Placebo?

The central theme of this book is that much – if not all – of the therapeutic effects of antidepressants are due to the placebo effect. Might this not also be true of the effect of psychotherapy on depression? Could this also be a placebo effect? This is one of the objections that I hear quite often when I am invited to speak about my research. Psychotherapy is no more effective than antidepressant medication, these critics contend. So if antidepressants are merely placebos, so too is psychotherapy.

If you look back again at the graph in Chapter 1 (page 10) showing the results of the first meta-analysis that I published on the treatment of depression, you can see why people might conclude that psychotherapy – like antidepressants – is merely a placebo. My own analysis showed that the effectiveness of psychotherapy is about the same as that of drugs, and that although both are much better than no treatment at all, neither is much better than placebo pills.[29]

In the short run, psychotherapy is about as effective as medication, which means that it is only slightly more effective than placebo pills. In the long run, however, CBT is much more effective than antidepressant drugs, which means that it is also much more effective than placebos. Still, there is a sense in which the critics are right. There is a good reason for thinking of the effects of psychotherapy as being similar to placebo effects, even though the research shows it to be more effective than placebo pills.

Dan Moerman, an anthropologist at the University of Michigan, has pointed out that the phrase 'placebo effect' is really an oxymoron, a contradiction in terms.[30] By definition, placebos are supposed to be inert. So how could they possibly affect anything? As a solution to this definitional conundrum, Moerman has coined the term 'meaning response' to designate what up till now we have called the placebo effect. Meanings are not inert. They can and do affect people. In fact, a fundamental premise of Albert Ellis's rational emotive therapy, which was the first cognitive therapy for emotional problems, is that the way we feel does not depend on the events that happen to us, but rather on the meaning these events have for us.

Imagine trying to design a research study to control for the meaning effect in psychotherapy. How could one ever do this? I suppose we might replace the meaningful words that the psychotherapist uses with similar-sounding gibberish. Perhaps we could have the therapist speak only in a language that the patient does not understand. But even then, some meaning would be assigned to the treatment, as when a priest or shaman chants in a sacred language or a doctor describes your condition and its treatment in convoluted technical jargon.

The point is that meaning is the essence of psychotherapy. It is through meaning that treatment effects are supposed to be brought about. Controlling for the meaning effect in psychotherapy is like controlling for the drug effect in the evaluation of a medicine. It just makes no sense.

Moerman's concept of the meaning effect shows the futility of trying to 'control for the placebo effect' in studies of psychotherapy. Nevertheless, researchers have devised a number of procedures that are intended to do just that. Most commonly, these are referred to as 'attention control' procedures. Their specific components vary greatly. In fact, some of the interventions that have been used as attention-control or placebo procedures have also been evaluated as bona-fide psychotherapies. What they have in common is that they are all supposed to control for the effects of being given attention and treatment by a clinician, which is a component of the placebo effect in medicine. These attention-control procedures differ in their effectiveness, but on the whole they are significantly less effective than real psychotherapies.[31] This is one reason for rejecting arguments that dismiss psychotherapy as merely a placebo.

Like antidepressants, a substantial part of the benefit of psychotherapy depends on the placebo effect, or as Moerman calls it, the meaning response. At least part of the improvement that is produced by these treatments is due to the relationship between the therapist and the client and to the client's expectancy of getting better. That is a problem for antidepressant treatment. It is a problem because drugs are supposed to work because of their chemistry, not because of psychological factors. But it is not a problem for psychotherapy. Psychotherapists are trained to provide a warm and caring environment in which therapeutic change can take place. Their intention is to replace the hopelessness of depression with a sense of hope and faith in the future.[32] These tasks are part of the essence of psychotherapy. The fact that psychotherapy can mobilize the meaning response – and that it can do so without deception – is one of its strengths, not one of its weaknesses. Because hopelessness is a fundamental characteristic of depression, instilling hope is a specific treatment for it. Invoking the meaning response is essential for the effective treatment of depression, and the best treatments are those that can do this most effectively and that can do so without deception.

As we have seen, the meaning response can be very large. In the treatment of depression, it is much larger than the drug effect. In fact, if you take away the meaning response, there may be no drug effect left at all. So what we need is a means of evoking this response. We want to exploit it rather than avoid it, and a treatment that can capitalize on the meaning response without deception should be embraced rather than rejected. What we need is a way to activate a therapeutic meaning response in clinical practice, and to do so without deceiving people or playing tricks on them by giving them sugar pills. That is exactly what psychotherapy is supposed to do, and that is what it does. That is why I call it the quintessential placebo.

The Costs of Psychotherapy

There is yet another advantage of psychotherapy, and it is one that is counter-intuitive. Psychotherapy costs less than antidepressants. At first glance it might seem impossible for this claim to be true. Certainly, a week's worth of antidepressants costs less than a 50-minute session of psychotherapy. Still, in the long run, psychotherapy is cheaper. Psychotherapy is cheaper because many patients have to remain on antidepressants if they are not to relapse and become depressed again. In contrast, all of the psychological treatments that have been tested and found effective in the long-term treatment of depression are relatively brief treatments that last from 10 to 20 weeks. After that there are no additional costs. About nine months after the beginning of treatment, the costs of continuing antidepressant treatment catch up to the costs of brief psychotherapy, and after that, the cumulative costs of medication continue to rise, whereas those of psychotherapy do not.[33]

NICE has recognized the importance of psychotherapy in their current guidelines on the treatment of depression.[34] They recommend six to eight sessions of CBT or some other form of psychotherapy or counselling for mild or moderate depression, CBT for recurrent depression, and CBT combined with antidepressants for severe depression. The problem, of course, is resources. When doctors prescribe psychotherapy in the UK, patients generally have to wait from six to nine months for an NHS therapist. In some cases the patient may have to wait up to two years, and in some areas therapy may not be available at all. The lack of resources creates a dilemma for GPs and for their patients who are suffering from depression. Surveys indicate that most doctors would prescribe antidepressants less often if other treatment options were available without long waiting lists.[35]

Currently, the UK government is taking steps to make psychotherapy for depression more readily available. On 20 January 2005 the Prime Minister's Strategy Unit hosted a seminar in the Cabinet Office, the focus of which was an invited paper presented by Lord Richard Layard, entitled 'Mental Health: Britain's Biggest Health Problem'.[36] In his paper, Lord Layard argued forcefully for a ten-year plan in which 10,000 new therapists would be trained to provide CBT and other therapies that had been shown to be effective in clinical trials. According to Layard, the programme would not only pay for itself, but would actually generate a profit. Depression can lead to time-off at work, physical health problems and hospitalization. This costs society money in terms of reduced output of goods and services, and it costs the taxpayer money in terms of benefits, services and reduced tax revenue. The cost of short-term psychotherapy would be about £750, but the government would save £850 per patient in reduced incapacity benefits and higher taxes alone, not to mention the costs saved by the NHS through reduced medication prescriptions and hospitalization.

On 12 May 2005 the UK government launched the programme that Lord Layard had advocated by establishing two pilot centres, one in Doncaster and the other in Newham, where CBT would be offered as an alternative to medication to people suffering from depression. Two years later the pilot programme was deemed a success, and the government announced that it would be expanded with the development of ten new 'Pathfinder' sites.[37] Large numbers of people, including patients who had applied without being referred by a GP, had been treated in a short time frame. Recovery rates were consistent with the clinical trials I have reviewed in this chapter, and statutory sick pay was reduced.

If the government's Pathfinder programme is a success, the problem of insufficient therapists may be solved. But what do we do in the meantime? People who are depressed cannot wait until the year 2015 for help. Fortunately, there are some low-cost alternatives that are available right now. These are treatment approaches that are sometimes used in conjunction with psychotherapy, but can also be used as stand-alone treatments. Let us take a look at them.

ST JOHN'S WORT

St John's wort is a yellow flowering plant that was first used medicinally by the ancient Greeks as a diuretic and a treatment for wounds and menstrual disorders. This herbal remedy is widely prescribed in Germany, where it has been studied extensively in clinical trials as a treatment for depression. In most countries, including the UK, it is available over the counter. In Ireland it is available only by prescription. Recently, a team of German scientists led by Klaus Linde at the University of Munich published a comprehensive review of 29 clinical trials of St John's wort, involving more than 5,000 depressed patients. They concluded that it is more effective than placebos and as effective as standard antidepressants in the treatment of major depression.

To be fair, since conventional antidepressants are not much better than placebos, one would have to draw the same conclusion about St John's wort. Still, it has some advantages over standard antidepressants. In particular, it generates far fewer side effects. In fact, the percentage of patients reporting side effects on St John's wort does not seem to be significantly more than the percentage from placebos.[38] To me, this means that the difference in effectiveness between St John's wort and placebos, while small, may be more genuine than the difference between conventional anti-depressants and placebos. Recall that the effectiveness of regular antidepressants in clinical trials is linked to the side effects they produce. Side effects are a cue that enables patients to 'break blind' and realize that they have been given the real drug. This can produce an enhanced placebo effect, which is responsible for at least part of the difference between drug and placebo. Because St John's wort does not have appreciably more side effects than placebos, patients in clinical trials of this herbal remedy are much less likely to break blind. As a result, the benefit that it shows compared to placebos may be more trustworthy than the equally small benefit of antidepressant drugs.

There are, however, some drawbacks to St John's wort. In countries where it is sold over the counter, there may be a lack of government oversight over its production, but this would be easy to remedy. The most important drawback is that it affects the way the body processes a number of other drugs, including conventional antidepressants and birth-control pills. Like any other drug, St John's wort would need to be taken under consultation with a doctor who knows what other medications the patient is taking.

The reaction to St John's wort by the medical profession reveals an interesting double standard. For example, a large clinical trial sponsored by the National Institutes of Health in the United States has been interpreted as showing that it is 'no more effective than placebo in treating major depression'.[39] In fact, the clinical trial on which this conclusion was based included a group of patients that was given the SSRI Seroxat.[40] Although St John's wort did not do significantly better than placebos in that trial, neither did Seroxat. So if this trial shows that St John's wort does not work, it also shows that antidepressants don't work. Nevertheless, it is often cited as evidence against St John's wort, but not against SSRIs.

I am not a great fan of St John's wort. For lasting control of depression, psychological treatment produces the best results, and medication does not add much – if anything – to it. Nevertheless, if a depressed patient wants medication, or if available alternative treatments are not sufficiently effective, this herbal remedy, taken under medical guidance, may be worth considering.

PHYSICAL EXERCISE

Physical exercise as a treatment for clinical depression has not been studied as extensively as drugs or psychotherapy, but there are a number of clinical trials evaluating its effectiveness.[41] In some of these studies, exercise was compared to no treatment at all. In others, it was compared to psychotherapy, medication or attention-control procedures intended to control for the non-specific placebo aspects of the exercise programme. Some of the trials also looked at the combination of physical exercise with medication, to see if the two treatments together might be more effective than either one alone. The results of these studies have been summarized nicely in an official 2004 report for the NHS by Sir Liam Donaldson, the Chief Medical Officer for England. Sir Liam concluded that 'physical activity is effective in the treatment of clinical depression and can be as successful as psychotherapy or medication, particularly in the longer term'.[42]

The studies of physical exercise as a treatment for depression contain a number of surprising findings. First, exercise is more effective for moderate to severe depression than it is for mild to moderate depression. Second, the antidepressant benefits of exercise seem to be long-lasting, so long as the person keeps exercising regularly. In fact, the benefits of exercise seem to increase as time goes on. Twenty minutes of exercise three days a week seems to be enough to produce the antidepressant effect, and the kind of exercise that is practised does not seem to matter much. Walking and running are equally effective, and anaerobic exercises like weight training are as effective as aerobic exercise. Finally, epidemiological studies indicate that exercise can prevent depression as well as ameliorate it.[43]

In 2000, a group of researchers led by James Blumenthal at Duke University in North Carolina reported the results of a particularly important clinical trial assessing exercise as a treatment for depression.[44] Equally depressed patients were randomly divided into three groups. One group was given a four-month course of aerobic exercise, the SSRI Lustral was prescribed to a second group, and the third group was given both Lustral and the exercise course. After four months of treatment there were no significant differences between the groups. Patients in all three groups had improved significantly. In other words, exercise was as effective as medication in lowering depression, and combining the two treatments was no more effective than using just one of them.

But the most interesting findings from this clinical trial were obtained six months later – ten months after the beginning of the study. Some important differences between the three treatment groups emerged at this follow-up assessment. By this point, significantly more exercise patients had recovered from depression, and more SSRI patients had relapsed. In other words, exercise was more effective than drugs.

This advantage of physical exercise over medication in the long run is reminiscent of comparisons between cognitive behavioural psychotherapy and antidepressants. More people treated with antidepressants relapse than those treated with either CBT or physical exercise. But there is an interesting difference between the psychotherapy data and the exercise data. Not only did the patients who had been assigned to the exercise group fare better than those in the drug group, but they also did better than those in the combined exercise plus medication group. In other words, adding an SSRI to exercise training increased the risk of getting depressed again. This was something that Blumenthal and his colleagues had not anticipated when they designed their study. They had assumed that if combining exercise with medication had any effect at all, it would be a positive one, in which the two treatments together would be more effective than either of them alone.

How can we explain this rather strange finding that exercise alone was more helpful than exercise combined with antidepressants? The drugs seemed to have had a harmful effect, somehow making the exercise programme less effective. This is consistent with comments that were made by some of the people who were in the group that combined exercise with drugs. According to the researchers, a number of the patients in this group 'mentioned spontaneously that the medication seemed to interfere with the beneficial effects of the exercise program'. But how did the medication achieve this negative effect? One possibility is that it was a nocebo effect. People may have volunteered for this study because it offered an alternative to drug treatment, and, in fact, some of the participants expressed disappointment when they were told they would be given an antidepressant drug. Their negative feelings about the drug component of treatment may have blunted the positive effect of the exercise programme.

There seems to be considerable reluctance in some parts of the medical community to acknowledge the benefits of exercise in the treatment of depression. One meta-analysis of clinical trials showed that physical exercise was as effective as psychotherapy or antidepressant medication and much better than no treatment. But the authors concluded that 'the effectiveness of exercise in reducing symptoms of depression cannot be determined',[45] and the editors of the journal introduced the article with an editorial comment entitled 'effectiveness of exercise in managing depression is not shown by meta-analysis'.[46] Why not? Because there were flaws in the way many of the studies had been designed. To be fair, there were indeed shortcomings in the studies, but these shortcomings also characterize clinical trials of antidepressants.[47] If clinical trials like these do not establish the effectiveness of physical exercise as a treatment for depression, neither do they establish the effectiveness of antidepressants.

How does physical exercise alleviate depression? One possibility is that it increases the release of endorphins that produce a sense of well-being, sometimes referred to as the 'runner's high'. Another possibility is that it is a placebo effect. But even if it is a placebo effect, consider the differences between exercise and antidepressants in side effects. Side effects of antidepressants include sexual dysfunction, nausea, vomiting, insomnia, drowsiness, seizures, diarrhoea and headaches. Side effects of physical exercise include enhanced libido, better sleep, decreased body fat, improved muscle tone, greater life expectancy, increased strength and endurance and improved cholesterol levels. So if both antidepressants and exercise work by means of the placebo effect, which placebo would you prefer?

If physical exercise is as effective as psychotherapy, why bother with psychotherapy at all? Why not just prescribe exercise? It is true that exercise is as effective as psychotherapy when all forms of psychotherapy are lumped together, but when different types of psychotherapy are compared with exercise, a somewhat different picture emerges.[48] Exercise is about as effective as psychodynamic therapy and more effective than supportive counselling, but it is less effective than CBT. The next step will be to assess the effectiveness of combining CBT with physical-exercise programmes. That might turn out to be the most effective treatment of all. One of my hopes is that a researcher reading this book will conduct a clinical trial to find out if this hypothesis is right.

PSYCHOTHERAPY WITHOUT PSYCHOTHERAPISTS

Many of the benefits of CBT can be obtained without going into therapy. There are a number of self-help books, CDs and computer programs that have been used to treat depression and some of these have been tested in clinical trials with positive results. I can particularly recommend two of these books. One is *Control Your Depression*, the lead author of which is Peter Lewinsohn, a Professor of Psychology at the University of Oregon.[49] Beginning in the 1970s, Lewinsohn pioneered the use of behaviour therapy for the treatment of depression, and the treatment procedures that he and his colleagues proposed have since become standard components of CBT. The other book that I can recommend with confidence is *Feeling Good* by the psychiatrist David Burns.[50] Burns based his approach on the cognitive-therapy programme developed by Aaron Beck at the University of Pennsylvania. This is the type of psychotherapy that is most often meant when the term CBT is used. *Control Your Depression* emphasizes behavioural techniques like increasing pleasant activities, improving social skills and learning to relax. *Feeling Good* puts greater emphasis on changing the way people think about themselves. But both books include both cognitive and behavioural techniques.

As a psychotherapist, I have recommended both of these books to depressed clients, and I found them useful adjuncts to treatment, but the real basis of my recommendation is the research that has been published testing their effectiveness as stand-alone treatments for depression. You might wonder whether something so simple as reading a book could possibly cure depression, but clinical studies indicate that it can. An analysis of these trials shows that people get less depressed after reading these books, and a three-year follow-up indicates that the benefits are long-lasting.[51]

The most prolific researcher of bibliotherapy is Forrest Scogin, a Professor of Psychology at the University of Alabama. One of Scogin's studies compared the clinical effectiveness of *Feeling Good* to standard CBT with a live therapist. Although patients in both groups improved, those who had seen a therapist had improved more than the others by the end of treatment. But the subjects who had been given the book to read continued to improve,and within three months they had caught up with those who had received standard CBT. One caveat is needed, however. The patients studied in clinical trials of bibliotherapy were only moderately depressed. We do not yet know what effect books like *Feeling Good* and *Control Your Depression* would have on people who are more severely depressed, but for those who are mildly or moderately depressed, working through the exercises in these books can be a reasonable alternative to psychotherapy.

Physicians might wonder whether self-help treatments would be acceptable to their depressed patients. A recent study by Alastair Dobbin, a general practitioner in Edinburgh, suggests that they might actually prefer it.[52] Dobbin let depressed patients referred by the NHS choose between taking antidepressant medication prescribed by their GPs and receiving a self-help self-hypnosis treatment programme presented on CDs. Eighty-six per cent of the patients chose the self-help self-hypnosis programme, 7 per cent chose antidepressants and the rest expressed no preference. With so few patients in the drug condition, a statistical comparison of the outcomes of the two treatments was not feasible, but those getting the self-help programme did at least as well as patients in studies of antidepressants and CBT.

SOCIAL CHANGE

In this chapter I have stressed the good news that there are many effective treatments for combating depression. This is necessarily true given the strength of the placebo effect. If placebos produce improvement, then any credible bona-fide treatment will also alleviate depression. Some of these treatments may be more effective than placebos, but in the treatment of depression, the placebo effect is always a major component.

That's the good news. Unfortunately, there is also some bad news. The bad news is that despite the range of treatments available, many people remain depressed after treatment and others relapse after getting better. Even CBT, which can substantially reduce the likelihood of relapse, does not eliminate it altogether.

Depression is not just an individual problem; it is also a social problem. The people most likely to become depressed are poor, unemployed and undereducated.[53] To some extent, this may be due to what is called social selection or economic drift. People who are chronically depressed might find it harder to perform well or even hold a job, and this might lead to a downward shift in their economic status. But there are data showing that the cause and effect can also run in the opposite direction.[54] Different ethnic groups, for example, have different rates of depression. As the authors of one of the studies investigating this pointed out, 'ethnic status cannot be an effect of disorder because it is present at birth'. Another study showed that people are more likely to become depressed if their parents were poor or less educated. These data cannot be explained by the economic-drift hypothesis. In other words, poverty and discrimination can cause depression.

The importance of economic problems in depression has been shown in a study of psychotherapy for depression conducted by two researchers in Chicago.[55] They found that during the first two sessions of treatment, more than 85 per cent of the depressed patients spontaneously brought up issues relating to inadequate financial resources, difficult working conditions or unemployment. They also found that the patients did better if their therapists responded by focusing on their economic problems as part of the treatment.

Still, dealing effectively with depression requires more than merely treating it. Not only are poor, unemployed, less well-educated and non-white people more likely to become depressed, but they are also least likely to benefit from treatment by either antidepressants or psychotherapy.[56] That is why combating depression requires more than merely providing effective treatment for those who are already suffering from it. We also need to change the social conditions – such as racism, unemployment, poverty, unaffordable housing and lack of adequate education – that put people at increased risk of becoming depressed.

Using data collected by the World Health Organization, Richard Wilkinson and Kate Pickett have shown that countries in which there is greater economic inequality have higher rates of mental illness. Their conclusions were based on data from rich countries only, ranging from the more economically equal Japan and Belgium to the less equal US and UK. So it was not the level of poverty per se that made the difference. Rather it was the unequal distribution of income within each country that was associated with emotional disorders and other social problems.[57] These data reinforce the idea that decreasing social inequality might also reduce the incidence of depression.

Epilogue

By now, I hope that I have convinced you that much of what has passed for common wisdom about depression and antidepressants is simply wrong. Depression is not caused by a chemical imbalance in the brain, and it is not cured by medication. Depression may not even be an illness at all. Often, it can be a normal reaction to abnormal situations. Poverty, unemployment, and the loss of loved ones can make people depressed, and these social and situational causes of depression cannot be changed by drugs.

Depression is a serious problem, but drugs are not the answer. In the long run, psychotherapy is both cheaper and more effective, even for very serious levels of depression. Physical exercise and self-help books based on CBT can also be useful, either alone or in combination with therapy. Reducing social and economic inequality would also reduce the incidence of depression.

'DON'T ROCK THE BOAT'

Reporting these conclusions and the evidence on which they are based has not always been an easy task. I have faced some rather hostile crowds at medical schools, although more often the reception has been open and cordial. Nevertheless, there can be negative consequences to taking a stance that challenges powerful interests. I recall working with a researcher at a medical school some years ago in an effort to design a clinical study of antidepressants using the balanced placebo design that I described in Chapter 3. Our collaboration ended when he was warned that he should not submit a grant proposal with my name on it, if he ever wanted to do a clinical trial on antidepressants again. I cannot blame him for this decision, as he was funded on 'soft money', which means that his salary depended on getting his research funded.

I was well established in my career when this incident happened. Having a tenured position in the psychology department at the University of Connecticut, I did not feel threatened by it. In fact, I must admit to feeling a bit proud of my apparent infamy. But young researchers with their careers at stake are also subject to this sort of pressure. One young colleague on the staff at a medical school wrote a paper critical of antidepressants that was published in a very distinguished medical journal. Instead of being proud of him, his department head told him that he should not have written the article, that he should not become too involved with me and that he was biting the hand that feeds him.

In August 2000, David Healy was formally offered a position at the Centre for Addiction and Mental Health, a teaching hospital affiliated with the University of Toronto. Three months later Healy gave an invited address at the university, during which he noted that most of the clinical trials of Lustral and Prozac had 'failed to detect any treatment effect'. This claim is actually much milder than it seems at first glance. Healy concluded that these unsuccessful trials did not constitute evidence that the drugs were ineffective. Instead, like many of the critics of my recent work, he saw them as 'evidence of the inadequacy of our assessment methods'.[1] On the other hand, Healy did say that he believed that SSRIs can lead to suicide, and in subsequent years he has backed that claim up with persuasive evidence.[2] In any case, one week after delivering his lecture, Healy received a message withdrawing the offer of the post at the hospital.

The most recent 'don't rock the boat' incident that I am aware of involved Jonathan Leo, Associate Professor of Neuroanatomy at Lincoln Memorial University in Tennessee. Leo and his colleague Jeffrey Lacasse, an assistant professor in the school of social work at Arizona State University, had criticized an article that had been published in the *Journal of the American Medical Association* (*JAMA*). The study supported the use of the SSRI Cipralex (Lexapro in the US) to prevent depression in patients who have suffered a stroke. It also assessed the effects of problem-solving therapy, a form of CBT, and found that it too prevented depression. But until Lacasse and Leo's questioning, the authors of the study had not directly compared the two forms of treatment. When they did, they found the two to be equally effective.[3]

Lacasse and Leo's criticism, and the reply to it, were published in *JAMA* in October 2008. Five months later Leo and Lacasse wrote another commentary on the *JAMA* article and sent it to the *British Medical Journal*. They recounted the story of having wondered about the comparison between Lexapro and problem-solving therapy and noted the results of the comparison that had been published in response to their questioning. They also mentioned an apparent conflict of interest involving financial connections between Robert Robinson, the lead author of the study, and Forest Pharmaceuticals, the US manufacturer of Lexapro. Robinson acknowledged the conflict of interest and apologized for not having disclosed it.[4]

All of this is pretty standard stuff. Researchers write articles. Other researchers criticize them, sometimes vociferously, often in the same journal, but sometimes in other journals. I have already lost count of the number of challenging commentaries that my most recent meta-analysis provoked in journals other than the one in which my article had been published. Sometimes I was alerted to them by editors, most often not. Occasionally I was invited to reply.

What happened to Jonathan Leo next was reported in the *Wall Street Journal*'s online Health Blog. *JAMA* editors phoned Leo and the Dean of his institution. According to Leo, the deputy editor of *JAMA* asked him, 'Who do you think you are', and then said,

'You are banned from *JAMA* for life. You will be sorry. Your school will be sorry. Your students will be sorry.' The *JAMA* editors confirmed the calls. They said that Leo had exaggerated their content, but the editor-in-chief of the journal is also quoted as telling the *Wall Street Journal* reporter, 'This guy is a nobody and a nothing. He is trying to make a name for himself. He should be spending time with his students instead of doing this.'[5] The editor subsequently denied making such a statement, claiming the journalist had misquoted her. Nevertheless, the general response to Leo is scary stuff from one of the world's leading medical journals.

'DON'T ASK, DON'T TELL'

When Bill Clinton was campaigning for the presidency of the United States, he promised to lift the long-standing ban on homosexuals serving in the military. But once in office, congressional opposition was so strong that he was forced to back off. The result was a compromise, in which gays were allowed to remain in the armed forces, as long as they did not call attention to their sexual orientation. This compromise came to be known as the 'Don't ask, don't tell' policy.

Sometimes, when they run out of arguments in defence of antidepressants, people suggest that I should have adopted a 'Don't ask, don't tell' policy. Even if the drugs don't work, they tell me, it is wrong to say so in public or to write about it in medical-journal articles, like the ones my colleagues and I have published. They argue that we shouldn't tell patients that the drugs don't work, even if it is true, because it will undermine their faith in treatment.

In 2004 the FDA urged drug companies to adopt a 'Don't ask, don't tell' policy with respect to their clinical-trial data showing that antidepressants are not better than placebos for depressed children. If the data were made public, they cautioned, it might lead doctors to not prescribe antidepressants. The FDA believed that the jury was still out on antidepressants for children. Even if the clinical trials show negative results, an FDA spokesperson was reported to have said to a *Washington Post* reporter, it doesn't mean that the drugs are ineffective.[6] The assumption seems to have been that doctors should prescribe medications that have not been shown to work, until it has been proven that they don't work.

I disagree strongly with the 'Don't ask, don't tell' policy. Without accurate knowledge, patients and physicians cannot make informed treatment decisions, researchers will ask the wrong questions and policymakers will implement misinformed policies. If the antidepressant effect is largely or entirely a placebo effect, it is important that we know this. If placebos can make people better, then depression can be ameliorated without reliance on drugs that have potentially serious side effects and that foster dependency.

For people who are depressed, and especially for those who do not receive enough benefit from medication or for whom the side effects of antidepressants are troubling, the fact that placebos can duplicate much of the effects of antidepressants should be taken as good news. It means that there are other ways of alleviating depression. As we have seen, treatments like psychotherapy and physical exercise are at least as effective as antidepressant drugs and more effective than placebos. In particular, CBT has been shown to lower the risk of relapsing into depression for years after treatment has ended, making it particularly cost-effective.

For society as a whole, knowledge of what the data on antidepressants really say should be a clarion call. Resources need to be made available for the provision of effective alternative treatments, and the social and economic causes of depression need to be addressed and overcome. It is my hope that this book will contribute to a wider recognition of the need for these changes in public policy and attitudes.

As you may have gathered by now, I rather enjoy telling tales and ruffling feathers. I also enjoy rocking boats, especially when they are in need of sinking. I hope you have enjoyed the ride.

Notes

Preface

1 John P. A. Ioannidis, 2008; Jeffrey Lacasse and Jonathan Leo, 2005; 'CNS Drug Discoveries: What the Future Holds 2008'.
2 Irving Kirsch, 1990.
3 John D. Teasdale, 1985.
4 Irving Kirsch and Guy Sapirstein, 1998.
5 Irving Kirsch, Alan Scoboria and Thomas J. Moore, 2002b; Irving Kirsch, Thomas J. Moore et al., 2002a; Irving Kirsch, Brett J. Deacon et al., 2008.
6 NICE, 'Depression: Management of Depression in Primary and Secondary Care'; CSIP, Choice and Access Programme, 2007; Eero Castrén, 2005; H. G. Ruhé, N. S. Mason and Aart H. Schene, 2007.

1 Listening to Prozac, but Hearing Placebo

1 William Schofield, 1964.
2 Standardized mean difference between pre-treatment and post-treatment depression scores for each type of treatment, Irving Kirsch and Guy Sapirstein, 1998.
3 John P. A. Ioannidis, 2008.
4 John W. Williams, Jr, et al., 2000.
5 Joanna Moncrieff, 2008b.
6 Mark S. Kramer et al., 1998.
7 Michael Philipp, Ralf Kohnen and Karl O. Hiller, 1999.
8 Carl Sherman, 1998.

9 J. G. Rabkin et al., 1986.
10 Joel R. Sneed et al., 2008; Martin Enserink, 1999; Martin Keller et al., 2006.
11 Roger P. Greenberg et al., 1994.
12 James M. Ferguson, 2001.
13 Greenberg et al., 1994.
14 John F. Kihlstrom, 1998.
15 Corrado Barbui, Toshiaki A. Furukawa and Andrea Cipriani, 2008.
16 Corrado Barbui, Andrea Cipriani and Irving Kirsch, 2009.
17 Joanna Moncrieff, S. Wessely and R. Hardy, 2005; Joanna Moncrieff, 2008b.

2 The 'Dirty Little Secret'

1 Peter Nathan and Martin E. P. Seligman, 1998.
2 Larry E. Beutler, 1998; Donald F. Klein, 1998.
3 Russell Joffe, Stephen Sokolov and David Streiner, 1996.
4 Hans Melander et al., 2003.
5 Irving Kirsch and Guy Sapirstein, 1998; Richard A. Hansen et al., 2005; Gerald Gartlehner et al., 2007.
6 Irving Kirsch, Thomas J. Moore et al., 2002a; Irving Kirsch, Brett J. Deacon et al., 2008.
7 NICE, 'Depression: Management of Depression in Primary and Secondary Care'.
8 Ibid.
9 Kirsch, Deacon et al., 2008.
10 Anton J. M. de Craen et al., 1999.
11 Carl Sherman, 1998.
12 Otto Benkert et al., 1997.
13 David O. Antonuccio, David D. Burns and William G. Danton, 2002; Roger P. Greenberg, 2002; Walter A. Brown, 2002; Michael E. Thase, 2002.
14 Steven D. Hollon et al., 2002.
15 Melander et al., 2003.
16 Wayne Kondro and Barbara Sibbald, 2004; 'Major Pharmaceutical Firm Concealed Drug Information', 2004.
17 Kondro and Sibbald, 2004.

18 Martin Keller, Neal D. Ryan et al., 2001.
19 'Major Pharmaceutical Firm Concealed Drug Information', 2004.
20 Gardiner Harris, 2004.
21 Alex Berenson, 2005.
22 B. J. Deacon, Kimberlee Glassner and Irving Kirsch, 2006; Melander et al., 2003.
23 Melander et al., 2003.
24 NICE, 'Depression: Management of Depression in Primary and Secondary Care'.
25 Catherine DeAngelis, Jeffrey M. Drazen et al., 2004.
26 Paul Leber, 1998.
27 Shankar Vedantam, 2004.
28 Jerry Avorn, 2007.
29 Ibid.
30 EMEA, 2008; Rob Evans and Sarah Boseley, 2004.
31 Hans Melander, Tomas Salmonson et al., 2008.
32 Irving Kirsch and Joanna Moncrieff, 2007.
33 Thomas P. Laughren, 1998.
34 Leber, 1998.
35 Laughren, 1998.

3 Countering the Critics

1 'Doctors Change Prescribing Habits on Back of SSRI Study', 2008.
2 'Antidepressants Work . . . ,' David Nutt, quoted in Martin Enserink, 2008; 'Dozens of Clinical Trials,' Rachel Werner, 2008.
3 Arthur K. Shapiro and L. A. Morris, 1978.
4 Arthur K. Shapiro, 1960.
5 Joel R. Sneed et al., 2008.
6 Lene Vase, Joseph L. Riley III and Donald D. Price, 2002.
7 Madhukar H. Trivedi et al., 2006; A. John Rush et al., 2006a; A. John Rush et al., 2006b.
8 Blair T. Johnson and Irving Kirsch, 2008.
9 S. Wolf et al., 1957.
10 Robert E. Kelly, Jr, et al., 2006.
11 Matthew J. Taylor et al., 2006.
12 Joseph Glenmullen, 2006; Christopher H. Warner et al., 2006.

13 H. G. Ruhé, N. S. Mason and Aart H. Schene, 2007; Giovanni A. Fava, 2003.
14 Arif Khan, Nick Redding and Walter A. Brown, 2008.
15 Hypericum Depression Trial Study Group, 2002.
16 James L. Claghorn and John P. Feighner, 1993.
17 Khan, Redding, and Brown, 2008.
18 NICE, 'Depression: Management of Depression in Primary and Secondary Care'.
19 Erick H. Turner et al., 2008.
20 Quoted in Marilyn Elias, 2002.
21 J. G. Rabkin et al., 1986.
22 Ted J. Kaptchuk, 1998b.
23 Sandra Lee et al., 2004.
24 Editorial, 'A Double-Edged Sword', 2008. I should also note that *Nature*, which is the primary journal of the Nature Publishing Group, responded to our meta-analysis with an excellent editorial on 6 March 2008. Citing the difficulties we had in obtaining access to complete data, they advocated a mandatory database that would provide access to the results of all trials clinical trials that are undertaken, not just those that are published.
25 Hans Melander, Tomas Salmonson et al., 2008.
26 D. S. Charney et al., 2002; Michael A. Posternak et al., 2002.
27 Trivedi et al., 2006.
28 Sneed et al., 2008.
29 Werner, 2008.
30 Corrado Barbui, Andrea Cipriani and Irving Kirsch, 2009.
31 GlaxoSmithKline, 2008.
32 Corrado Barbui, Toshiaki A. Furukawa and Andrea Cipriani, 2008.
33 Irving Kirsch, 2000.
34 G. A. Marlatt and D. J. Rohsenow, 1980.
35 I. Kirsch and M. J. Rosadino, 1993; Fabrizio Benedetti, G. Maggi et al., 2003.
36 Sneed et al., 2008.
37 Irving Kirsch, 'Are drug and placebo effects in depression additive?' *Biological Psychiatry*, 47, 733-735, 2000.
38 Pedro L. Delgado, 2000.

4 The Myth of the Chemical Imbalance

1 Jeffrey R. Lacasse and Jonathan Leo, 2005; P. J. Cowen, 2008.
2 David Healy, 1997; Joanna Moncrieff, 2008a.
3 Francisco López-Muñoz et al., 2007.
4 Peter Fangmann et al., 2008.
5 Roland Kuhn, 1958.
6 Healy, 1997.
7 Roland Kuhn, 1990.
8 Joseph J. Schildkraut, 1965.
9 Alec Coppen, 1967.
10 Healy, 1997.
11 López-Muñoz et al., 2007; Healy, 1997.
12 Schildkraut, 1965.
13 Julius Axelrod, L. G. Whitby and George Hertting, 1961; Julius Axelrod and Joseph K. Inscoe, 1963.
14 Schildkraut, 1965.
15 F. K. Goodwin and W. E. Bunney, Jr, 1971.
16 A. John Rush et al., 2006a.
17 D. L. Davies and Michael Shepherd, 1955.
18 Michael Shepherd, 1956.
19 Healy, 1997.
20 Axelrod, Whitby and Hertting, 1961.
21 Schildkraut, 1965.
22 Lacasse and Leo, 2005; Joanna Moncrieff, 2008b; Eero Castrén, 2005.
23 Joseph Mendels and Alan Frazer, 1974.
24 H. G. Ruhé, N. S. Mason and Aart H. Schene, 2007.
25 Ibid. For the sake of clarity, I have altered the quotation by spelling out some of the abbreviations.
26 Coppen, 1967.
27 G. S. Malhi, G. B. Parker and J. Greenwood, 2005.
28 Andrea Cipriani et al., 2009. The calculations are simple and straightforward. Table 3 of *The Lancet* article reports response rates for head-to-head comparisons of different antidepressants, along with the number of subjects on which each response rate was based. I merely extracted the response rates in all of the head-to-head comparisons of an SSRI with an NDRI, multiplied each response rate by the number of subjects it was based on, summed the product and divided

the sum by the total number of subjects.

29 A. John Rush et al., 2006b.
30 Gerald Gartlehner et al., 2007; Richard A. Hansen et al., 2005; Cipriani et al., 2009.
31 Hansen et al., 2005.
32 Robert E. Kelly, Jr, et al., 2006.
33 Hansen et al., 2005.
34 Rush et al., 2006b.
35 Moncrieff, 2008b; Irving Kirsch and Guy Sapirstein, 1998; Irving Kirsch, 2003.
36 Sheldon H. Preskorn, 2004; Milan Sarek, 2006.
37 Siegfried Kasper and Bruce S. McEwen, 2008; Antona J. Wagstaff, Douglas Ormrod and Caroline M. Spencer, 2001; Tayfun I. Uzbay, 2008.
38 Wagstaff, Ormrod and Spencer, 2001.
39 Thomas Kuhn, 1970.
40 I. Hindmarch, 2002.
41 Castrén, 2005.
42 Ibid.
43 Michael E. Hyland, 1985.
44 Helen S. Mayberg, Mario Liotti et al., 1999; Helen S. Mayberg, J. Arturo Silva et al., 2002.

5 The Placebo Effect and the Power of Belief

1 Jeremy Laurance, 2008; 'Depression Drugs Don't Work, Says New Study', 2008; Sarah Boseley, 2008; Fiona McRae, 2008.
2 Rebecca Smith, 2008.
3 Jeff Aronson, 1999; Geoffrey Chaucer, 2003.
4 T. C. Graves, 1920.
5 'The Humble Humbug', 1954.
6 Ted J. Kaptchuk, Catherine E. Kerr and Abby Zanger, 2009.
7 Alfred Binet and Charles Féré, 1988.
8 Ted J. Kaptchuk, 1998a; Ted J. Kaptchuk 1998c.
9 S. Wolf, 1950.
10 Ibid.; F. K. Abbot, M. Mack and S. Wolf, 1952.
11 H. K. Beecher, 1955.
12 E. F. Traut and E. W. Passarelli, 1957.

13 Kaptchuk, 1998c.
14 A. Hróbjartsson and P. C. Gøtzsche, 2001; A. Hróbjartsson and P. C. Gøtzsche, 2004.
15 Lene Vase, Joseph L. Riley III and Donald D. Price, 2002; Joel R. Sneed et al., 2008.
16 Anton J. M. de Craen, D. E. Moerman et al., 1999; Anton J. M. de Craen, J. G. Tijssen et al., 2000; A. Branthwaite and P. Cooper, 1981; Rebecca L. Waber et al., 2008.
17 Vase, Riley III and Price, 2002.
18 Peter Tyrer et al., 2008.
19 De Craen, Tijssen et al., 2000; Ted J. Kaptchuk, W. B. Stason et al., 2006.
20 Christopher G. Goetz et al., 2008.
21 Mario Battezzati, Alberto Tagliaferro and Angelo Domenko Cattaneo, 1959.
22 L. Cobb et al., 1959; E. G. Dimond, C. F. Kittle and J. E. Crockett, 1960.
23 Dimond, Kittle and Crockett, 1960.
24 Traut and Passarelli, 1957.
25 Margaret Talbot, 2000.
26 J. Bruce Moseley et al., 2002.
27 Talbot, 2000.
28 David F. Felson and Joseph Buckwalter, 2002.
29 Nelda P. Wray, J. Bruce Moseley and K. O'Malley, 2002.
30 Alexandra Kirkley et al., 2008.
31 Robert G. Marx, 2008.
32 Irving Kirsch, 1990.
33 Michael E. Hyland, 1985; Irving Kirsch and Michael E. Hyland, 1987.
34 For more complete discussions of the relation between mind and brain, see Peter M. Churchland, 1984.
35 Helen S. Mayberg, Maria Liotti et al., 1999; Helen S. Mayberg, Steven K. Brannon et al., 2000; Helen S. Mayberg, J. Arturo Silva et al., 2002.
36 Mayberg, Silva et al., 2002; p. 731.
37 Kimberly Goldapple et al., 2004; Andrew F. Leuchter et al., 2002.
38 Sarah-Jayne Blakemore and Uta Frith, 2005.
39 Tor D. Wager, James K. Rilling et al., 2004.
40 Samantha C. Sodergren and Michael E. Hyland, 1999.

41 T. J. Luparello, H. A. Lyons et al., 1968.
42 T. J. Luparello, N. Leist, et al., 1970.
43 Y. Ikemi and S. Nakagawa, 1962.
44 B. Klopfer, 1957.
45 Per-Henrik Zahl, Jan Mæhlen and H. Gilbert Welch, 2008.
46 Michael D. Storms and Richard E. Nisbett, 1970.
47 Timothy F. Jones et al., 2000.
48 William Lorber, Giuliana Mazzoni and Irving Kirsch, 2007.
49 A. M. Daniels and R. Sallie, 1981.
50 M. G. Myers, J. A. Calms and J. Singer, 1987.
51 Roy R. Reeves et al., 2007.
52 W. B. Cannon, 1942.
53 Esther M. Sternberg, 2002.
54 Michael Philipp, Ralf Kohnen and Karl O. Hiller, 1999; Corrado Barbui, Andrea Cipriani and Irving Kirsch, 2009.
55 Irving Kirsch, 1985; Irving Kirsch (ed.), 1999.
56 S. Reiss and R. J. McNally, 1985; Kirsch, 1985.
57 John D. Teasdale, 1985.

6 How Placebos Work

1 Zelda Di Blasi et al., 2001; Ted J. Kaptchuk, John M. Kelley et al., 2008.
2 Kaptchuk, Kelley et al., 2008.
3 Ted J. Kaptchuk, 1983; Ted J. Kaptchuk, 2000.
4 Anton J. M. de Craen, D. E. Moerman et al., 1999.
5 Scot H. Simpson et al., 2006.
6 Kaptchuk, Kelley et al., 2008.
7 Di Blasi et al., 2001; Kaptchuk, Kelley et al., 2008.
8 Roger S. Ulrich, 1984.
9 Margaret A. Chesney et al., 2005.
10 D. Räikkönen et al., 1999; D. M. Byrnes et al., 1998; M. F. Scheier et al., 1999; David Spiegel and Janine Giese-Davis, 2003; H. Yang and W. Lin, 2005.
11 Irving Kirsch, 2006.
12 R. Pogge, 1963.
13 Irving Kirsch and Lynne J. Weixel, 1988; M. Frankenhaeuser et al., 1963.

14 Irving Kirsch, 1985; Irving Kirsch (ed.), 1999; Steve Stewart-Williams and John Podd, 2004.

15 Arthur K. Shapiro, E. Struening and E. Shapiro, 1980.

16 Guy H. Montgomery and Irving Kirsch, 1996.

17 Donald D. Price and Howard L. Fields, 1997.

18 Jon D. Levine, Newton C. Grodon and Howard L. Fields, 1978.

19 Tor D. Wager, David J. Scott and Jon-Kar Zubieta, 2007.

20 Fabrizio Benedetti, C. Arduino and M. Amanzio, 1999.

21 Kirsch, 1985; Stewart-Williams and Podd, 2004; Tor D. Wager, 2005; Tor D. Wager, James K. Rilling et al., 2004.

22 Ivan P. Pavlov, 'Physiology of Digestion: Nobel Lecture 12 December 1904; Ivan P. Pavlov, 1927; Robert E. Clark, 2004.

23 N. J. Voudouris, C. L. Peck and G. Coleman, 1985; N. J. Voudouris, C. L. Peck and G. Coleman, 1989; N. J. Voudouris, C. L. Peck and G. Coleman, 1990.

24 Donald D. Price, Leonard B. Milling et al., 1999.

25 Wager, Rilling et al., 2004.

26 Robert A. Rescorla, 1988; Irving Kirsch, Steven J. Lynn et al., 2004; Voudouris, Peck and Coleman, 1985.

27 Voudouris, Peck and Coleman, 1985.

28 Guy H. Montgomery and Irving Kirsch, 1997.

29 Alison Watson et al., 2007.

30 Fabrizio Benedetti, Antonella Pollo et al., 2003.

31 Irene Elkin, 1994.

32 Stuart M. Sotsky et al., 1991.

33 Irving Kirsch, 1990; Joel Weinberger and Andrew Eig, 1999; Björn Meyer et al., 2002.

34 Aaron T. Beck et al., 1979; John D. Teasdale, 1985.

35 Frederic M. Quitkin et al., 1998.

36 Aimee M. Hunter et al., 2006.

37 Kirsch and Weixel, 1988.

38 Joel R. Sneed et al., 2008.

7 Beyond Antidepressants

1 Irving Kirsch and Guy Sapirstein, 1998.

2 Joel R. Sneed et al., 2008.

3 R. Hamish McAllister-Williams, 2008.

4 Richard A. Hansen et al., 2005; Michael Philipp, Ralf Kohnen and Karl O. Hiller, 1999; Corrado Barbui, Andrea Cipriani and Irving Kirsch, 2009.
5 FDA, 2006.
6 Bettina C. Prator, 2006.
7 Tarek A. Hammad, Thomas Laughren and Judith Racoosin, 2006.
8 Marc B. Stone and M. Lisa Jones, 2006; David Healy, 2009; Dean Fergusson et al., 2005.
9 Peter R. Breggin, 2003/2004; Peter R. Breggin, 2006; David Healy, Andrew Herxheimer and David B. Menkes, 2006.
10 Anthony J. Rothschild and Carol A. Locke, 1991.
11 Christopher H. Warner et al., 2006; Joseph Glenmullen, 2006.
12 Jon C. Tilburt et al., 2008; Uriel Nitzan and Pesach Lichtenberg, 2004.
13 Tilburt et al., 2008.
14 Lee C. Park and Lino Covi, 1965.
15 Fabrizio Benedetti, Antonella Pollo, et al., 2003.
16 Andrew C. Butler et al., 2006.
17 NICE, 'Depression: Management of Depression in Primary and Secondary Care'; Claudi L. H. Bockting et al., 2005; Keith S. Dobson et al., 2008; Giovanni A. Fava et al., 2004.
18 Zac E. Imel et al., 2008.
19 Peter M. Lewinsohn, David O. Antonuccio et al., 1984; Aaron T. Beck et al., 1979.
20 Gerald L. Klerman et al., 1984.
21 H. Davanloo, 1976.
22 Carl Rogers, 1961.
23 Pim Cuijpers et al., 2008; Sona Dimidjian et al., 2006; Dobson et al., 2008; Leslie A. Robinson, Jeffrey S. Berman and Robert A. Neimeyer, 1990.
24 Stone and Jones, 2006; Hammad, Laughren and Racoosin, 2006; FDA, 2007.
25 NICE, 'Depression: Management of Depression in Primary and Secondary Care'.
26 Bockting et al., 2005.
27 Fava et al., 2004.
28 Michael A. Friedman et al., 2004; Marc B. J. Blom et al., 2007.
29 Kirsch and Sapirstein, 1998.

30 Daniel E. Moerman, 2006; Daniel E. Moerman and Wayne B. Jonas, 2002.
31 Bruce E. Wampold et al., 2002.
32 J. D. Frank, 1973.
33 Dobson et al., 2008.
34 NICE, 'Depression: Management of Depression in Primary and Secondary Care'.
35 Ed Halliwell, 2005.
36 Richard Layard, 2004; Richard Layard, 2006.
37 CSIP, Choice and Access Programme, 2007a; CSIP, 2007b; CSIP, 2008.
38 Philipp, Kohnen and Hiller, 1999.
39 NCAM, 'St John's Wort'.
40 Hypericum Depression Trial Study Group, 2002.
41 Lynette L. Craft and Daniel M. Landers, 1998; Debbie A. Lawlor and Stephen W. Hopker, 2001; James A. Blumenthal et al., 1999; Michael A. Babyak et al., 2000; Nalin A. Singh, Karen M. Clements and Maria A. Fiatarone Singh, 2001.
42 Liam Donaldson, 2004.
43 William J. Strawbridge et al., 2002.
44 Blumenthal et al., 1999; Babyak et al., 2000.
45 Lawlor and Hopker, 2001.
46 Editorial, 'Effectiveness of Exercise', 2001.
47 M. Hotopf, G. Lewis and C. Normand, 1997.
48 M. Pinquart, P. M. Duberstein and J. M. Lyness, 2007.
49 Peter M. Lewinsohn, R. F. Munoz et al., 1978.
50 David D. Burns, 1980.
51 Robert J. Gregory et al., 2004; Mark Floyd et al., 2004; Nancy M. Smith et al., 1997.
52 Alastair Dobbin, Margaret Maxwell and Robert Elton, 2009.
53 Christopher G. Hudson, 2005; V. Lorant et al., 2003.
54 Bruce P. Dohrenwend et al., 1992; Jeffrey G. Johnson et al., 1999.
55 Lydia Falconnier and Irene Elkin, 2008.
56 Madhukar H. Trivedi et al., 2006; R. Bruce Sloane et al., 1976.
57 Richard Wilkinson and Kate Pickett, 2009.

Epilogue

1 David Healy, 'Psychopharmacology and the Government of the Self'.
2 David Healy 2003; Dean Fergusson et al., 2005; David Healy, 2004.
3 Robert G. Robinson et al., 2008; Jeffrey Lacasse and Jonathan Leo, 2008.
4 Jonathan Leo and Jeffrey Lacasse, 2009; Sharon Davies, 2009.
5 David Armstrong, 2009; see also the *JAMA* editors' reply: Catherine D. DeAngelis and Phil B. Fontanarosa, 2009, and Leo's response to it: Jonathan Leo, 2009.
6 Shankar Vedantam, 2004.

Bibliography

Abbot, F. K., M. Mack and S. Wolf, 'The Action of Banthine on the Stomach and Duodenum of Man with Observations of the Effects of Placebos', *Gastroenterology* 20 (1952)

Antonuccio, David O., David D. Burns and William G. Danton, 'Antidepressants: A Triumph of Marketing over Science?', *Prevention & Treatment*, no. 25 (2002); http://www.journals.apa.org/prevention/volume5/pre0050025c.html

Armstrong, David, 'Jama Editor Calls Critic a "Nobody and a Nothing"', in *Wall Street Journal Health Blog*, 2009

Aronson, Jeff, 'When I Use a Word . . . Please, Please Me', *British Medical Journal* 318 (1999): 716

Avorn, Jerry, 'Paying for Drug Approvals – Who's Using Whom?', *New England Journal of Medicine* 356 (2007): 1697–700

Axelrod, Julius and Joseph K. Inscoe, 'The Uptake and Binding of Circulating Serotonin and the Effect of Drugs', *Journal of Pharmacology and Experimental Therapeutics* 141, no. 2 (1963): 161–65

——, L. G. Whitby and George Hertting, 'Effect of Psychotropic Drugs on the Uptake of H^3-Norepinephrine by Tissues', *Science* 133, no. 3450 (1961): 383–84

Babyak, Michael A., James A. Blumenthal, Steve Herman, Parinda Khatri, P. Murali Doraiswamy, Kathleen A. Moore, W. Edward Craighead, Teri T. Baldewicz and K. Ranga Krishnan, 'Exercise Treatment for Major Depression: Maintenance of Therapeutic Benefit at 10 Months', *Psychosomatic Medicine* 62 (2000): 633–38

Barbui, Corrado, Andrea Cipriani and Irving Kirsch, 'Is the Paroxetine–Placebo Efficacy Separation Mediated by Adverse Events? A Systematic Re-Examination of Randomised Double-Blind Studies', submitted for publication (2009)

——, Toshiaki A. Furukawa and Andrea Cipriani, 'Effectiveness of

Paroxetine in the Treatment of Acute Major Depression in Adults: A Systematic Re-Examination of Published and Unpublished Data from Randomized Trials', *Canadian Medical Association Journal* 178, no. 3 (2008): 296–305

Battezzati, Mario, Alberto Tagliaferro and Angelo Domenko Cattaneo, 'Clinical Evaluation of Bilateral Internal Mammary Artery Ligation as Treatment of Coronary Heart Disease', *American Journal of Cardiology* 4 (1959): 180–83

Beck, Aaron T., A. J. Rush, B. F. Shaw and G. Emery, *Cognitive Therapy of Depression*, New York: Guilford, 1979

Beecher, H. K., 'The Powerful Placebo', *Journal of the American Medical Association* 159, no. 17 (1955): 1602–06

Benedetti, Fabrizio, C. Arduino and M. Amanzio, 'Somatotopic Activation of Opioid Systems by Target-Directed Expectations of Analgesia', *Journal of Neuroscience* 19 (1999): 3639–48

——, G. Maggi, L. Lopiano, M. Lanotte, I. Rainero, S. Vighetti and A. Pollo, 'Open Versus Hidden Medical Treatments: The Patient's Knowledge About a Therapy Affects the Therapy Outcome', *Prevention & Treatment* (2003); http://journals.apa.org/prevention/volume6/pre0060001a.html

——, Antonella Pollo, Leonardo Lopiano, Michele Lanotte, Sergio Vighetti and Innocenzo Rainero, 'Conscious Expectation and Unconscious Conditioning in Analgesic, Motor, and Hormonal Placebo/Nocebo Responses', *Journal of Neuroscience* 23, no. 10 (2003): 4315–23

Benkert, Otto, A. Szegedi, H. Wetzel, H. J. Staab, W. Meister and M. Philipp, 'Dose Escalation Vs. Continued Doses of Paroxetine and Maprotiline: A Prospective Study in Depressed out-Patients with Inadequate Treatment Response', *Acta Psychiatrica Scandinavica* 95 (1997): 288–96

Berenson, Alex, 'Despite Vow, Drug Makers Still Withhold Data', *New York Times*, 31 May 2005

Beutler, Larry E., 'Prozac and Placebo: There's a Pony in There Somewhere', *Prevention & Treatment*, Article 0003c (1998); http://journals.apa.org/prevention/volume1/pre0010003c.html

Binet, Alfred and Charles Féré, *Animal Magnetism*, New York: Appleton, 1888

Blakemore, Sarah-Jayne and Uta Frith, *The Learning Brain: Lessons for Education*, Malden, MA: Blackwell Publishing, 2005

Blom, Marc B. J., Kosse Jonker, Elise Dusseldorp, Philip Spinhoven, Erik Hoencamp, Judith Haffmans and Richard van Dyck,

'Combination Treatment for Acute Depression Is Superior Only When Psychotherapy Is Added to Medication', *Psychotherapy and Psychosomatics* 76 (2007): 289–97

Blumenthal, James A., Michael A. Babyak, Kathleen A. Moore, W. Edward Craighead, Steve Herman, Parinda Khatri, Robert Waugh, Melissa A. Napolitano, Leslie M. Forman, Mark Appelbaum, P. Murali Doraiswamy and K. Ranga Krishnan, 'Effects of Exercise Training on Older Patients with Major Depression', *Archives of Internal Medicine* 159 (1999): 2349–56

Bockting, Claudi L. H., Aart H. Schene, Philip Spinhoven, Maarten W. J. Koeter, Luuk F. Wouters, Jochanan Huyser, Jan H. Kamphuis and The DELTA Study Group, 'Preventing Relapse/Recurrence in Recurrent Depression with Cognitive Therapy: A Randomized Controlled Trial', *Journal of Consulting & Clinical Psychology* 73, no. 4 (2005): 647–57

Boseley, Sarah, 'Prozac, Used by 40m People, Does Not Work Say Scientists', *Guardian*, 26 February 2008

Branthwaite, A. and P. Cooper, 'Analgesic Effects of Branding in Treatment of Headaches', *British Medical Journal* (*Clin Res Ed*) 282, no. 6276 (1981): 1576–78

Breggin, Peter R., 'How Glaxosmithkline Suppressed Data on Paxil-Induced Akathisia: Implications for Suicidality and Violence', *Ethical Human Psychology and Psychiatry: An International Journal of Critical Inquiry* 8, no. 2 (2006): 91–100

——, 'Suicidality, Violence and Mania Caused by Selective Serotonin Reuptake Inhibitors (SSRIs): A Review and Analysis', *International Journal of Risk & Safety in Medicine* 16 (2003/2004): 31–49

Brown, Walter A., 'Are Antidepressants as Ineffective as They Look?', *Prevention & Treatment*, no. 26 (2002); http://www.journals.apa.org/prevention/volume5/pre0050026c.html

Burns, David D., *Feeling Good: The New Mood Therapy*, New York: Avon Books, 1980

Butler, Andrew C., Jason E. Chapman, Evan M. Forman and Aaron T. Beck, 'The Empirical Status of Cognitive-Behavioral Therapy: A Review of Meta-Analyses', *Clinical Psychology Review* 26 (2006): 17–31

Byrnes, D. M., M. H. Antoni, K. Goodkin, J. Efantis-Potter, D. Asthana, T. Simon et al., 'Stressful Events, Pessimism, Natural Killer Cell Cytotoxicity, and Cytotoxic/Suppressor T Cells in Hiv+ Black Women at Risk for Cervical Cancer', *Psychosomatic Medicine* 60 (1998): 714–22

Cannon, W. B., '"Voodoo" Death', *American Anthropologist* 44 (1942): 169–81

Castrén, Eero, 'Is Mood Chemistry?', *Nature Reviews Neuroscience* 6 (2005): 241–46

Charney, D. S., C. B. Nemeroff, L. Lewis, S. K. Laden, J. M. Gorman, E. M. Laska, M. Borenstein, C. L. Bowden, A. Caplan, G. J. Emslie, D. L. Evans, B. Geller, L. E. Grabowski, J. Herson, N. H. Kalin, P. E. Keck, I. Kirsch, K. R. R. Krishnan, D. J. Kupfer, R. W. Makuch, F. G. Miller, H. Pardes, R. Post, M. M. Reynolds, L. Roberts, J. F. Rosenbaum, D. L. Rosenstein, D. R. Rubinow, A. J. Rush, N. D. Ryan, G. S. Sachs, A. F. Schatzberg and S. Solomon, 'National Depressive and Manic-Depressive Association Consensus Statement on the Use of Placebo in Clinical Trials of Mood Disorders', *Archives of General Psychiatry* 59, no. 3 (2002): 262–70

Chaucer, Geoffrey, 'The Persones Tale,' in *Complete Works of Geoffrey Chaucer, Part 2*, edited by Walter W. Skeat, Whitefish, MT: Kessinger Publishing, 2003, pp. 675–717

Chesney, Margaret A., Lynae A. Darbes, Kate Hoerster, Jonelle M. Taylor, Donald B. Chambers and David E. Anderson, 'Positive Emotions: Exploring the Other Hemisphere in Behavioral Medicine', *International Journal of Behavioral Medicine* 12, no. 2 (2005): 50–58

Churchland, Peter M., *Matter and Consciousness*, Cambridge, MA: MIT Press, 1984

Cipriani, Andrea, Toshiaki A. Furukawa, Georgia Salanti, John R. Geddes, Julian P. T. Higgins, Rachel Churchill, Norio Watanabe, Atsuo S. Nakagawa, Ichiro M. Omori, Hugh McGuire, Michele Tansella and Corrado Barbui, 'Comparative Efficacy and Acceptability of 12 New Generation Antidepressants: A Multiple Treatments Meta-Analysis, *The Lancet* (2009)

Claghorn, James L. and John P. Feighner, 'A Double-Blind Comparison of Paroxetine with Imipramine in the Long-Term Treatment of Depression', *Journal of Clinical Psychopharmacology* 13 (1993): 23S–27S

Clark, Robert E., 'The Classical Origins of Pavlov's Conditioning', *Integrative Physiological & Behavioral Science* 39, no. 4 (2004): 279–94

'CNS Drug Discoveries: What the Future Holds 2008'; http://www.marketwatch.com/news/story/cns-drug-discoveries-future-holds/story.aspx?guid=%7BFD07856B-48A9-496A-95F6-74155042DADE%7D&dist=hppr

Cobb, L., G. I. Thomas, D. H. Dillard, K. A. Merendino and R. A. Bruce, 'An Evaluation of Internal-Mammary Artery Ligation by a

Double Blind Technique', *New England Journal of Medicine* 260 (1959): 1115–18

Coppen, Alec, 'The Biochemistry of Affective Disorders', *British Journal of Psychiatry* 113 (1967): 1237–64

Cowen, P. J., 'Serotonin and Depression: Pathophysiological Mechanism or Marketing Myth?', *Trends Pharmacol Sci*, no. 9 (2008)

Craft, Lynette L. and M. Daniel Landers, 'The Effects of Exercise on Clinical Depression and Depression Resulting from Mental Illness: A Metaregression Analysis', *Journal of Sport and Exercise Psychology* 20 (1998): 339–57

CSIP, Choice and Access Programme, 'Commissioning a Brighter Future: Improving Access to Psychological Therapies', edited by Department of Health: Crown, 2007a

——, 'Improving Access to Psychological Therapies: Specification for the Commissioner-Led Pathfinder Programme,' edited by Department of Health: Crown, 2007b

——, 'Improving Access to Psychological Therapies (Iapt): Commissioning Toolkit', edited by Department of Health: Crown, 2008

Cuijpers, Pim, Annemieke van Straten, Gerhard Andersson and Patricia van Oppen, 'Psychotherapy for Depression in Adults: A Meta-Analysis of Comparative Outcome Studies', *Journal of Consulting & Clinical Psychology* 76, no. 6 (2008): 909–22

Daniels, A. M. and R. Sallie, 'Headache, Lumbar Puncture, and Expectation', *The Lancet* 1, no. 8227 (1981): 1003

Davanloo, H., *Basic Principles and Techniques in Short-Term Depression*, New York: S. P. Medical & Scientific Books, 1976

Davies, D. L. and Michael Shepherd, 'Reserpine in the Treatment of Anxious and Depressed Patients', *The Lancet* 266, no. 6881 (1955): 117–20

Davies, Sharon, 'Potential Conflicts of Interest: More Information from Jama', *British Medical Journal* (2009); http://www.bmj.com/cgi/eletters/338/feb05_1/b463

Deacon, B. J., Kimberlee Glassner and Irving Kirsch, 'Publication Bias in Clinical Trials of SSRI Medications for Depression', conference presentation, Association for Behavioural and Cognitive Therapies (2006)

DeAngelis, Catherine, Jeffrey M. Drazen, Frank A. Frizelle, Charlotte Haug, John Hoey, Richard Horton, Sheldon Kotzin, Christine Laine, Ana Marusic, A. John, P. M. Overbeke, Torben V. Schroeder, Hal C. Sox and Martin B. Van Der Weyden, 'Clinical Trial Registration: A Statement from the International Committee of

Medical Journal Editors', *New England Journal of Medicine* 351, no. 12 (2004): 1250–51

DeAngelis, Catherine D. and Phil B. Fontanarosa, 'Conflicts over Conflicts of Interest', *Journal of the American Medical Association* (2009); http://jama.ama-assn.org/misc/jed90012pap_E1_E3.pdf

de Craen, Anton J. M., D. E. Moerman, S. H. Heisterkamp, G. N. J. Tytgat, J. G. Tijssen and J. Kleijnen, 'Placebo Effect in the Treatment of Duodenal Ulcer', *British Journal of Clinical Pharmacology* 48 (1999): 853–60

——, J. G. Tijssen, J. de Gans and J. Kleijnen, 'Placebo Effect in the Acute Treatment of Migraine: Subcutaneous Placebos Are Better Than Oral Placebos', *Journal of Neurology* 247 (2000): 183–88

Delgado, Pedro L., 'Depression: The Case for a Monoamine Deficiency', *Journal of Clinical Psychiatry* 61 (2000): 7–11

'Depression Drugs Don't Work, Says New Study', *The Times*, 26 February 2008

Di Blasi, Zelda, Elaine Harkness, Edzard Ernst, Amanda Georgioud and Jos Kleijnen, 'Influence of Context Effects on Health Outcomes: A Systematic Review', *The Lancet* 357, no. 9258 (2001): 757–62

Dimidjian, Sona, Steven D. Hollon, Keith S. Dobson, Karen B. Schmaling, Robert J. Kohlenberg, Michael E. Addis, Robert Gallop, Joseph B. McGlinchey, David K. Markley, Jackie K. Gollan, David C. Atkins, David L. Dunner and Neil S. Jacobson, 'Randomized Trial of Behavioral Activation, Cognitive Therapy, and Antidepressant Medication in the Acute Treatment of Adults with Major Depression', *Journal of Consulting and Clinical Psychology* 74, no. 4 (2006): 658–70

Dimond, E. G., C. F. Kittle and J. E. Crockett, 'Comparison of Internal Mammary Ligation and Sham Operation for Angina Pectoris', *American Journal of Cardiology* 5 (1960): 483–86

Dobbin, Alastair, Margaret Maxwell and Robert Elton, 'A Benchmarked Feasibility Study of a Self-Hypnosis Treatment for Depression in Primary Care', *International Journal of Clinical & Experimental Hypnosis* 57, no. 3 (2009): 293-318.

Dobson, Keith S., Steven D. Hollon, Sona Dimidjian, Karen B. Schmaling, Robert J. Kohlenberg, Robert J. Gallop, Shireen L. Rizvi, Jackie K. Gollan, David L. Dunner and Neil S. Jacobson, 'Randomized Trial of Behavioral Activation, Cognitive Therapy, and Antidepressant Medication in the Prevention of Relapse and Recurrence in Major Depression', *Journal of Consulting and Clinical Psychology* 76, no. 3 (2008): 468–77

'Doctors Change Prescribing Habits on Back of SSRI Study', *Onmedica News* (2008); http://www.onmedica.com/NewsArticle.aspx?id=ae98220c-10e5-4350-8a9b-c85d534

Dohrenwend, Bruce P., Itzhak Levav, Patrick E. Shrout, Sharon Schwartz, Guedalia Naveh, Bruce G. Link, Andrew E. Skodol and Ann Stueve, 'Socioeconomic Status and Psychiatric Disorders: The Causation-Selection Issue', *Science* 255, no. 5047 (1992): 946–52

Donaldson, Liam, 'At Least Five a Week: Evidence on the Impact of Physical Activity and Its Relationship to Health. A Report from the Chief Medical Officer', Department of Health, 2004

Editorial, 'A Double-Edged Sword', *Nature Reviews Drug Discovery* 7 (2008): 275

Editorial, 'Effectiveness of Exercise in Managing Depression Is Not Shown by Meta-Analysis', *British Medical Journal* 322 (2001)

Editorial, 'No More Scavenger Hunts', *Nature* (2008) 452, no. 7183:1

Elias, Marilyn, 'Study: Antidepressant Barely Better Than Placebo', *USA Today*, 7 July 2002

Elkin, Irene, 'The NIMH Treatment of Depression Collaborative Research Program: Where We Began and Where We Are', in *Handbook of Psychotherapy and Behavior Change*, edited by A. E. Bergin and S. L. Garfield, New York: Wiley, 1994, pp. 114–39

EMEA, 'Annual Report of the European Medicines Agency 2007', London: European Medicines Agency, 2008

Enserink, Martin, 'Can the Placebo Be the Cure?', *Science* 284 (1999): 238–40

——, 'The Problem with Prozac', *ScienceNOW Daily News*, 27 February 2008

Evans, Rob and Sarah Boseley, 'The Drugs Industry and Its Watchdog: A Relationship Too Close for Comfort?', *The Guardian*, 4 October 2004

Falconnier, Lydia and Irene Elkin, 'Addressing Economic Stress in the Treatment of Depression', *American Journal of Orthopsychiatry* 78, no. 1 (2008): 37–46

Fangmann, Peter, Hans-Jörg Assion, Georg Juckel, Cecilio Álamo González and Francisco López-Muñoz, 'Half a Century of Antidepressant Drugs: On the Clinical Introduction of Monoamine Oxidase Inhibitors, Tricyclics, and Tetracyclics. Part Ii: Tricyclics and Tetracyclics', *Journal of Clinical Psychopharmacology* 28, no. 1 (2008): 1–4

Fava, Giovanni A., 'Can Long-Term Treatment with Antidepressant Drugs Worsen the Course of Depression?', *Journal of Clinical Psychiatry* 64 (2003): 123–33

——, Chiara Ruini, Chiara Rafanelli, Livio Finos, Sandra Conti and Silvana Grandi, 'Six-Year Outcome of Cognitive Behavior Therapy for Prevention of Recurrent Depression', *American Journal of Psychiatry* 161 (2004): 1872–76

FDA, 'Combined Use of 5-Hydroxytryptamine Receptor Agonists (Triptans), Selective Serotonin Reuptake Inhibitors (SSRIs) or Selective Serotonin/Norepinephrine Reuptake Inhibitors (SNRIs) May Result in Life-Threatening Serotonin Syndrome', *FDA Public Health Advisory* (2006); http://www.fda.gov/Cder/Drug/advisory/SSRI_SS200607.htm

——, 'FDA Proposes New Warnings About Suicidal Thinking, Behavior in Young Adults Who Take Antidepressant Medications', Press release, 2007; http://www.fda.gov/bbs/topics/NEWS/2007/NEW01624.html

Felson, David F. and Joseph Buckwalter, 'Debridement and Lavage for Osteoarthritis of the Knee', *New England Journal of Medicine* 347 (2002): 132–33

Ferguson, James M., 'SSRI Antidepressant Medications: Adverse Effects and Tolerability', *The Primary Care Companion to The Journal of Clinical Psychiatry* 3, no. 1 (2001): 22–27

Fergusson, Dean, Steve Doucette, Kathleen Cranley Glass, Stan Shapiro, David Healy, Hebert Paul and Brian Hutton, 'Association between Suicide Attempts and Selective Serotonin Reuptake Inhibitors: Systematic Review of Randomised Controlled Trials', *British Medical Journal* 330 (2005): 396–99

Floyd, Mark, Forrest Scogin, Nancy L. McKendree-Smith, Donna L. Floyd and Paul D. Rokke, 'Cognitive Therapy for Depression: A Comparison of Individual Psychotherapy and Bibliotherapy for Depressed Older Adults', *Behavior Modification* 28 (2004): 297–318

Frank, J. D., *Persuasion and Healing*, revised ed., Baltimore: Johns Hopkins, 1973

Frankenhaeuser, M., G. Jarpe, H. Svan and B. Wrangsjö, 'Physiological Reactions to Two Different Placebo Treatments', *Scandinavian Journal of Psychology* 4 (1963): 245–50

Friedman, Michael A., Jerusha B. Detweiler-Bedell, Howard E. Leventhal, Rob Horne, Gabor I. Keitner and Ivan W. Miller, 'Combined Psychotherapy and Pharmacotherapy for the Treatment of Major Depressive Disorder', *Clinical Psychology: Science and Practice* 11, no. 1 (2004): 47–68

Gartlehner, Gerald, Richard A. Hansen, Patricia Thieda, Angela M. DeVeaugh-Geiss, Bradley N. Gaynes, Erin E. Krebs, Linda J. Lux,

Laura C. Morgan, Janelle A. Shumate, Loraine G. Monroe and Kathleen N. Lohr, 'Comparative Effectiveness of Second-Generation Antidepressants in the Pharmacologic Treatment of Adult Depression. Comparative Effectiveness Review No. 7. (Prepared by Rti International-University of North Carolina Evidence-Based Practice Center under Contract No. 290-02-0016.) Rockville, Md: Agency for Healthcare Research and Quality' (2007); www.effective-healthcare.ahrq.gov/reports/final.cfm

GlaxoSmithKline, 'Glaxosmithkline Statement: Public Library of Science Medicine Article on Antidepressant Medicines', London: Press release, 26 February 2008

Glenmullen, Joseph, *Coming Off Antidepressants*, London: Constable & Robinson, 2006

Goetz, Christopher G., Joanne Wuu, Michael P. McDermott, Charles H. Adler, Stanley Fahn, Curt R. Freed, Robert A. Hauser, Warren C. Olanow, Ira Shoulson, P. K. Tandon, Parkinson Study Group and Sue Leurgans, 'Placebo Response in Parkinson's Disease: Comparisons among 11 Trials Covering Medical and Surgical Interventions', *Movement Disorders* 5 (2008): 690–99

Goldapple, Kimberly, Zindel Segal, Carol Garson, Mark Lau, Peter Bieling, Sidney Kennedy and Helen Mayberg, 'Modulation of Cortical-Limbic Pathways in Major Depression: Treatment-Specific Effects of Cognitive Behavior Therapy', *Archives of General Psychiatry* 61, no. 1 (2004): 34–41

Goodwin, F. K. and W. E. Bunney, Jr, 'Depressions Following Reserpine: A Re-Evaluation', *Semin Psychiatry* 3, no. 435–48 (1971)

Graves, T. C., 'Commentary on a Case of Hystero-Epilepsy with Delayed Puberty: Treated with Testicular Extract', *The Lancet* 196 (1920): 1134–35

Greenberg, Roger P., 'Reflections on the Emperor's New Drugs', *Prevention & Treatment*, no. 27 (2002); http://www.journals.apa.org/prevention/volume5/pre0050027c.html

——, R. F. Bornstein, M. J. Zborowski, Seymour Fisher and M. D. Greenberg, 'A Meta-Analysis of Fluoxetine Outcome in the Treatment of Depression', *Journal of Nervous and Mental Disease* 182 (1994): 547–51

Gregory, Robert J., Sally Schwer Canning, Tracy W. Lee and Joan C. Wise, 'Cognitive Bibliotherapy for Depression: A Meta-Analysis', *Professional Psychology: Research and Practice* 35, no. 3 (2004): 275–80

Halliwell, Ed, *Up and Running*, London: Mental Health Foundation, 2005

Hammad, Tarek A., Thomas Laughren and Judith Racoosin, 'Suicidality in Pediatric Patients Treated with Antidepressant Drugs', *Archives of General Psychiatry* 63 (2006): 332–39

Hansen, Richard A., Gerald Gartlehner, Kathleen N. Lohr, Bradley N. Gaynes and Timothy S. Carey, 'Efficacy and Safety of Second-Generation Antidepressants in the Treatment of Major Depressive Disorder', *Annals of Internal Medicine* 143 (2005): 415–26

Harris, Gardiner, 'Maker of Paxil to Release All Trial Results', *New York Times*, 26 August 2004

Healy, David, *The Antidepressant Era*, Cambridge, MA: Harvard University Press, 1997

——, 'Are Selective Serotonin Reuptake Inhibitors a Risk Factor for Adolescent Suicide?', *Canadian Journal of Psychiatry* 54, no. 2 (2009): 69–71

——, *Let Them Eat Prozac: The Unhealthy Relationship between the Pharmaceutical Industry and Depression*, New York: New York University Press, 2004

——, 'Lines of Evidence on the Risks of Suicide with Selective Serotonin Reuptake Inhibitors', *Psychotherapy and Psychosomatics* 72, no. 2 (2003): 71–79

——, 'Psychopharmacology and the Government of the Self'; http://www.pharmapolitics.com/feb2healy.html

——, Andrew Herxheimer and David B. Menkes, 'Antidepressants and Violence: Problems at the Interface of Medicine and Law', *Public Library of Science Medicine (PLoS Med 3)*, no. 9 (2006): e372

Hindmarch, I., 'Beyond the Monoamine Hypothesis: Mechanisms, Molecules and Methods', *European Psychiatry* 17, Suppl 3 (2002): 294–99

Hollon, Steven D., Robert J. DeRubeis, Richard C. Shelton and Bahr Weiss, 'The Emperor's New Drugs: Effect Size and Moderation Effects', *Prevention & Treatment* 5, Article 27 (2002); http://www.journals.apa.org/prevention/volume5/pre0050027.html

Hotopf, M., G. Lewis and C. Normand, 'Putting Trials on Trial – the Costs and Consequences of Small Trials in Depression: A Systematic Review of Methodology', *Journal of Epidemiology and Community Health* 51 (1997): 354–58

Hróbjartsson, A. and P. C. Gøtzsche, 'An Analysis of Clinical Trials Comparing Placebo with No Treatment', *New England Journal of Medicine* 344 (2001): 1594–602

—— and P. E. Gøtzsche, 'Is the Placebo Powerless? Update of a Systematic Review with 52 New Randomized Trials Comparing

Placebo with No Treatment', *Journal of Internal Medicine* 256 (2004): 91–100

Hudson, Christopher G., 'Socioeconomic Status and Mental Illness: Tests of the Social Causation and Selection Hypotheses', *American Journal of Orthopsychiatry* 75, no. 1 (2005): 3–18

'The Humble Humbug', *The Lancet* 2 (1954): 321

Hunter, Aimee M., Andrew F. Leuchter, Melinda L. Morgan and Ian A. Cook, 'Changes in Brain Function (Quantitative EEG Cordance) During Placebo Lead-in and Treatment Outcomes in Clinical Trials for Major Depression', *American Journal of Psychiatry* 163, no. 8 (2006): 1426–32

Hyland, Michael E., 'Do Person Variables Exist in Different Ways?', *American Psychologist* 40 (1985): 1003–10

Hypericum Depression Trial Study Group, 'Effect of Hypericum Perforatum (St John's Wort) in Major Depressive Disorder: A Randomized Controlled Trial', *Journal of the American Medical Association* 287 (2002): 1807–14

Ikemi, Y. and S. Nakagawa, 'A Psychosomatic Study of Contagious Dermatitis', *Kyoshu Journal of Medical Science* 13 (1962): 335–50

Imel, Zac E., Melanie B. Malterer, Kevin M. McKay and Bruce E. Wampold, 'A Meta-Analysis of Psychotherapy and Medication in Unipolar Depression and Dysthymia', *Journal of Affective Disorders* 110 (2008): 197–206

Ioannidis, John P. A., 'Effectiveness of Antidepressants: An Evidence Myth Constructed from a Thousand Randomized Trials?', *Philosophy, Ethics, and Humanities in Medicine*, no. 14 (2008)

Joffe, Russell, Stephen Sokolov and David Streiner, 'Antidepressant Treatment of Depression: A Metaanalysis', *Canadian Journal of Psychiatry* 41 (1996): 613–16

Johnson, Blair T. and Irving Kirsch, 'Interpreting the Efficacy of Antidepressants: Statistical Significance Versus Clinical Benefits', *Significance* 5 (2008): 54–58

Johnson, Jeffrey G., Patricia Cohen, Bruce P. Dohrenwend, Bruce G. Link and Judith S. Brook, 'A Longitudinal Investigation of Social Causation and Social Selection Processes involved in the Association between Socioeconomic Status and Psychiatric Disorders', *Journal of Abnormal Psychology* 108, no. 3 (1999): 490–99

Jones, Timothy F., Allen S. Craig, Debbie Hoy, Elaine W. Gunter, David L. Ashley, Dana B. Barr, John W. Brock and William Schaffner, 'Mass Psychogenic Illness Attributed to Toxic Exposure at a High School', *New England Journal of Medicine* 342, no. 2 (2000): 96–100

Kaptchuk, Ted J., 'Intentional Ignorance: A History of Blind Assessment and Placebo Controls in Medicine', *Bulletin of the History of Medicine* 72, no. 3 (1998a): 389–433

——, 'Letter to the Editor', *Journal of the American Medical Association* 279, no. 19 (1998b): 1526–27

——, 'Powerful Placebo: The Dark Side of the Randomised Controlled Trial', *The Lancet* 351 (1998c): 1722–25

——, *The Web That Has No Weaver: Understanding Chinese Medicine*, New York: Congdon & Weed, 1983

——, *The Web That Has No Weaver: Understanding Chinese Medicine*, 2nd ed., New York: McGraw-Hill, 2000

——, John M. Kelley, Lisa A. Conboy, R. B. Davis, C. E. Kerr, E. E. Jacobson, Irving Kirsch, R. N. Schyner, B. Y. Nam, L. T. Nguyen, M. Park, A. L. Rivers, C. McManus, E. Kokkotou, D. A. Drossman, P. Goldman and A. J. Lembo, 'Components of the Placebo Effect: A Randomized Controlled Trial in Irritable Bowel Syndrome', *British Medical Journal* 336 (2008): 998–1003

——, Catherine E. Kerr and Abby Zanger, 'Placebo Controls, Exorcisms and the Devil', *The Lancet* 374 (2009): 1234–1235

——, W. B. Stason, R. B. Davis, A. T. R. Legedza, R. N. Schyner, C. E. Kerr, D. A. Stone, B. H. Nam, Irving Kirsch and R. H. Goldman, 'Sham Device Versus Inert Pill: A Randomized Controlled Trial Comparing Two Placebo Treatments for Arm Pain Due to Repetitive Use', *British Medical Journal* 332 (2006): 391–97

Kasper, Siegfried and Bruce S. McEwen, 'Neurobiological and Clinical Effects of the Antidepressant Tianeptine', *CNS Drugs* 22, no. 1 (2008): 15–26

Keller, Martin, Stuart Montgomery, William Ball, Mary Morrison, Duane Snavely, Guanghan Liu, Richard Hargreaves, Jarmo Hietala, Christopher Lines, Katherine Beebe and Scott Reines, 'Lack of Efficacy of the Substance P (Neurokinin1 Receptor) Antagonist Aprepitant in the Treatment of Major Depressive Disorder', *Biological Psychiatry* 59 (2006): 216–23

——, Neal D. Ryan, Michael Strober, Rachel G. Klein, Stan P. Kutcher, Boris Birmaher, Owen R. Hagino, Harold Koplewicz, Gabrielle A. Carlson, Gregory N. Clarke, Graham J. Emslie, David Feinberg, Barbara Geller, Vivek Kusumakar, George Papatheodorou, William H. Sack, Michael Sweeney, Karen Dineen Wagner, Elizabeth B. Weller, Nancy C. Winters, Rosemary Oakes and James P. McCafferty, 'Efficacy of Paroxetine in the Treatment of Adolescent Major

Depression: A Randomized, Controlled Trial', *Journal of the American Academy of Child & Adolescent Psychiatry* 40, no. 7 (2001): 762–72

Kelly, Jr, Robert E., Lisa J. Cohen, Randye J. Semple, Philip Bialer, Adam Lau, Alison Bodenheimer, Elana Neustadter, Arkady Barenboim and Igor I. Galynker, 'Relationship between Drug Company Funding and Outcomes of Clinical Psychiatric Research', *Psychological Medicine* 36, no. 1647–56 (2006)

Khan, Arif, Nick Redding and Walter A. Brown, 'The Persistence of the Placebo Response in Antidepressant Clinical Trials', *Journal of Psychiatric Research* 42, no. 10 (2008): 791–96

Kihlstrom, John F., 'Attributions, Awareness and Dissociation: In Memoriam Kenneth S. Bowers, 1937–1996', *American Journal of Clinical Hypnosis* 40, no. 3 (1998): 194–205

Kirkley, Alexandra, Trevor B. Birmingham, Robert B. Litchfield, J. Robert Giffin, Kevin R. Willits, Cindy J. Wong, Brian G. Feagan, Allan Donner, Sharon H. Griffin, Linda M. D'Ascanio, Janet E. Pope and Peter J. Fowler, 'A Randomized Trial of Arthroscopic Surgery for Osteoarthritis of the Knee', *New England Journal of Medicine* 359 (2008): 1097–107

Kirsch, Irving, 'Are Drug and Placebo Effects in Depression Additive?', *Biological Psychiatry* 47, no. 8 (2000): 733–35

——, *Changing Expectations: A Key to Effective Psychotherapy*, Belmont, CA: Brooks/Cole, 1990

—— (ed.), *How Expectancies Shape Experience*, Washington, DC: American Psychological Association, 1999

——, 'Placebo: The Role of Expectancies in the Generation and Alleviation of Illness', in *The Power of Belief: Psychosocial Influence on Illness, Disability and Medicine*, edited by Peter Halligan and Aylward Mansel, Oxford: Oxford University Press, 2006, pp. 55–67

——, 'Response Expectancy as a Determinant of Experience and Behavior', *American Psychologist* 40, no. 11 (1985): 1189–202

——, 'St John's Wort, Conventional Medication, and Placebo: An Egregious Double Standard', *Complementary Therapies in Medicine* 11, no. 3 (2003): 193–95

——, Brett J. Deacon, T. B. Huedo-Medina, Alan Scoboria, Thomas J. Moore and Blair T. Johnson, 'Initial Severity and Antidepressant Benefits: A Meta-Analysis of Data Submitted to the Food and Drug Administration', *PLoS Medicine*, no. 2 (2008); http://medicine.plosjournals.org/perlserv/?request=get-document&doi=10.1371/journal.pmed.0050045

—— and Michael E. Hyland, 'How Thoughts Affect the Body – a Metatheoretical Framework', *Journal of Mind and Behaviour* 8, no. 3 (1987): 417–34

——, Steven J. Lynn, M. Vigorito and R. R. Miller, 'The Role of Cognition in Classical and Operant Conditioning', *Journal of Clinical Psychology* 60, no. 4 (2004): 369–92

—— and Joanna Moncrieff, 'Clinical Trials and the Response Rate Illusion', *Contemporary Clinical Trials* 28, no. 4 (2007): 348–51

——, Thomas J. Moore, Alan Scoboria and Sarah S. Nicholls, 'The Emperor's New Drugs: An Analysis of Antidepressant Medication Data Submitted to the U.S. Food and Drug Administration', *Prevention & Treatment*, no. 23 (2002a); http://www.journals.apa.org/prevention/volume5/pre0050023a.html

—— and M. J. Rosadino, 'Do Double-Blind Studies with Informed Consent Yield Externally Valid Results – an Empirical-Test', *Psychopharmacology* 110, no. 4 (1993): 437–42

—— and Guy Sapirstein, 'Listening to Prozac but Hearing Placebo: A Meta-Analysis of Antidepressant Medication', *Prevention & Treatment*, Article 0002a (1998); http://www.journals.apa.org/prevention/volume1/pre0010002a.html

——, Alan Scoboria and Thomas J. Moore, 'Antidepressants and Placebos: Secrets, Revelations, and Unanswered Questions', *Prevention & Treatment* 5 (2002b): n.p.

—— and Lynne J. Weixel, 'Double-Blind Versus Deceptive Administration of a Placebo', *Behavioral Neuroscience* 102, no. 2 (1988): 319–23

Klein, Donald F., 'Listening to Meta-Analysis but Hearing Bias', *Prevention & Treatment*, Article 0006c (1998); http://journals.apa.org/prevention/volume1/pre0010006c.html

Klerman, Gerald L., Myrna M. Weissman, Bruce J. Rounsaville and Eve S. Chevron, *Interpersonal Psychotherapy of Depression*, New York: Basic Books, 1984

Klopfer, B., 'Psychological Variables in Human Cancer', *Journal of Projective Techniques* 21, no. 4 (1957): 331–40

Kondro, Wayne and Barbara Sibbald, 'Drug Company Experts Advised Staff to Withhold Data About SSRI Use in Children', *Canadian Medical Association Journal* 170, no. 5 (2004): 783

Kramer, Mark S., Neal Cutler, John Feighner, Ram Shrivastava, John Carman, John J. Sramek, Scott A. Reines, Guanghan Liu, Duane Snavely, Edwina Wyatt-Knowles, Jeffrey J. Hale, Sander G. Mills,

Malcolm MacCoss, Christopher J. Swain, Timothy Harrison, Raymond G. Hill, Franz Hefti, Edward M. Scolnick, Margaret A. Cascieri, Gary G. Chicchi, Sharon Sadowski, Angela R. Williams, Louise Hewson, David Smith, Emma J. Carlson, Richard J. Hargreaves and Nadia M. J. Rupniak, 'Distinct Mechanism for Antidepressant Activity by Blockade of Central Substance P Receptors', *Science* 281 (1998): 1640–45

Kuhn, Roland, 'Artistic Imagination and the Discovery of Antidepressants', *Journal of Psychopharmacology* 4, no. 3 (1990): 127–30

——, 'The Treatment of Depressive States with G 22355 (Imipramine Hydrochloride)', *American Journal of Psychiatry* 115, no. 5 (1958): 459–64

Kuhn, Thomas, *The Structure of Scientific Revolutions*, 2nd edn, Chicago: University of Chicago Press, 1970

Lacasse, Jeffrey and Jonathan Leo, 'Escitalopram, Problem-Solving Therapy, and Poststroke Depression', *Journal of the American Medical Association* 300, no. 15 (2008): 1757-b–58

—— and Jonathan Leo, 'Serotonin and Depression: A Disconnect between the Advertisements and the Scientific Literature', *PLoS Medicine* 2, no. 12 (2005): 1211–16

Laughren, Thomas P., 'Approvable Action on Forrest Laboratories, Inc. Nda 20–822 Celexa (Citalopram Hbr) for the Management of Depression', in *Memorandum to the Department of Health and Human Services, Public Health Service, Food and Drug Administration, Center for Drug Evaluation and Research*, Washington, DC, 26 March 1998

Laurance, Jeremy, 'Antidepressant Drugs Don't Work – Official Study', *The Independent*, 26 February 2008

Lawlor, Debbie A. and Stephen W. Hopker, 'The Effectiveness of Exercise as an Intervention in the Management of Depression: Systematic Review and Metaregression Analysis of Randomised Controlled Trials', *British Medical Journal* 322 (2001): 1–8

Layard, Richard, 'The Case for Psychological Treatment Centres', *British Medical Journal* 332 (2006): 1030–32

——, 'Mental Health: Britain's Biggest Health Problem', London: Prime Minister's Strategy Unit, 2004

Leber, Paul, 'Approvable Action on Forrest Laboratories, Inc. Nda 20–822 Celexa (Citalopram Hbr) for the Management of Depression', in *Memorandum to the Department of Health and Human Services, Public Health Service, Food and Drug Administration, Center for Drug Evaluation and Research*, Washington, DC, 4 May 1998

Lee, Sandra, John R. Walker, Laura Jakul and Kathryn Sexton, 'Does Elimination of Placebo Responders in a Placebo Run-in Increase the

Treatment Effect in Randomized Clinical Trials? A Meta-Analytic Evaluation', *Depression and Anxiety* 19 (2004): 10–19

Leo, Jonathan, 'Academic Freedom and Controversy over the Publication of Factually Correct, Publicly Available Information' (2009); http://online.wsj.com/public/resources/documents/leo_statement_for_WSJ.htm[25/04/2009 17:56:46]

—— and Jeffrey Lacasse, 'Clinical Trials of Therapy Versus Medication: Even in a Tie, Medication Wins', *British Medical Journal* (2009); http://www.bmj.com/cgi/eletters/338/feb05_1/b463

Leuchter, Andrew F., Ian A. Cook, Elise A. Witte, Melinda Morgan and Michelle Abrams, 'Changes in Brain Function of Depressed Subjects During Treatment with Placebo', *American Journal of Psychiatry* 159 (2002): 122–29

Levine, Jon D., Newton C. Grodon and Howard L. Fields, 'The Mechanism of Placebo Analgesia', *The Lancet* 2 (1978): 654–57

Lewinsohn, Peter M., David O. Antonuccio, J. S. Breckenridge and L. Teri, *The 'Coping with Depression' Course*, Eugene, OR: Castalia, 1984

——, R. F. Munoz, M. A. Youngren and A. M. Zeiss, *Control Your Depression*, New York: Macmillan Publishing, 1978

López-Muñoz, Francisco, Cecilio Álamo, Georg Juckel and Hans-Jörg Assion, 'Half a Century of Antidepressant Drugs: On the Clinical Introduction of Monoamine Oxidase Inhibitors, Tricyclics, and Tetracyclics. Part I: Monoamine Oxidase Inhibitors', *Journal of Clinical Psychopharmacology* 27, no. 6 (2007): 555–59

Lorant, V., D. Deliege, W. Eaton, A. Robert, P. Philippot and M. Ansseau, 'Socioeconomic Inequalities in Depression: A Meta-Analysis', *American Journal of Epidemiology* 157, no. 2 (2003): 98–112

Lorber, William, Giuliana Mazzoni and Irving Kirsch, 'Illness by Suggestion: Expectancy, Modeling, and Gender in the Production of Psychosomatic Symptoms', *Annals of Behavioral Medicine* 33 (2007): 112–16

Luparello, T. J., N. Leist, C. H. Lourie and P. Sweet, 'The Interaction of Psychologic Stimuli and Pharmacologic Agents on Airway Reactivity in Asthmatic Subjects', *Psychosomatic Medicine* 32, no. 5 (1970): 509–13

——, H. A. Lyons, E. R. Bleecker and E. R. McFadden, Jr, 'Influences of Suggestion on Airway Reactivity in Asthmatic Subjects', *Psychosomatic Medicine* 30, no. 6 (1968): 819–25

McAllister-Williams, R. Hamish, 'Misinterpretation of Randomized Trial Evidence: Do Antidepressants Work?', *British Journal of Hospital Medicine* 69, no. 5 (2008): 246–47

McRae, Fiona, 'Anti-Depressants Taken by Thousands of Brits "Do Not Work", Major New Study Reveals', *Daily Mail*, 26 February 2008

'Major Pharmaceutical Firm Concealed Drug Information: Glaxosmithkline Misled Doctors About the Safety of Drug Used to Treat Depression in Children', Press Release: Office of the New York State Attorney General Eliot Spitzer, 2 June 2004

Malhi, G. S., G. B. Parker and J. Greenwood, 'Structural and Functional Models of Depression: From Sub-Types to Substrates', *Acta Psychiatrica Scandinavica* 111, no. 2 (2005): 94–105

Marlatt, G. A., and D. J. Rohsenow, 'Cognitive Processes in Alcohol Use: Expectancy and the Balanced Placebo Design', in *Advances in Substance Abuse: Behavioral and Biological Research*, edited by N. K. Mello, Greenwich, CT: JAI Press, 1980, pp. 159–99

Marx, Robert G., 'Arthroscopic Surgery for Osteoarthritis of the Knee?', *New England Journal of Medicine* 259 (2008): 1169–70

Mayberg, Helen S., Steven K. Brannan, J. Janet L. Tekell, Arturo Silva, Roderick K. Mahurin, Scott McGinnis and Paul A. Jerabek, 'Regional Metabolic Effects of Fluoxetine in Major Depression: Serial Changes and Relationship to Clinical Response', *Biological Psychiatry* 48 (2000): 830–43

——, Mario Liotti, Stephen K. Brannan, Scott McGinnis, Roderick K. Mahurin, Paul A. Jerabek, J. Arturo Silva, J. Janet L. Tekell, Charles C. Martin, Jack L. Lancaster and Peter T. Fox, 'Reciprocal Limbic-Cortical Function and Negative Mood: Converging Pet Findings in Depression and Normal Sadness', *American Journal of Psychiatry* 156 (1999): 675–82

——, J. Arturo Silva, Steven K. Brannan, J. Janet L. Tekell, Roderick K. Mahurin, Scott McGinnis and Paul A. Jerabek, 'The Functional Neuroanatomy of the Placebo Effect', *American Journal of Psychiatry* 159 (2002): 728–37

Melander, Hans, Jane Ahlqvist-Rastad, Gertie Meijer and Björn Beermann, 'Evidence B(I)Ased Medicine – Selective Reporting from Studies Sponsored by Pharmaceutical Industry: Review of Studies in New Drug Applications', *British Medical Journal* 326 (2003): 1171–73

——, Tomas Salmonson, Eric Abadie and Barbara van Zwieten-Boot, 'A Regulatory Apologia – a Review of Placebo-Controlled Studies in Regulatory Submissions of New-Generation Antidepressants', *European Neuropsychopharmacology* (2008)

Mendels, Joseph and Alan Frazer, 'Brain Biogenic Amine Depletion and Mood', *Archives of General Psychiatry* 30, no. 4 (1974): 447–51

Meyer, Björn, Paul A. Pilkonis, Janice L. Krupnick, Matthew K. Egan, Samuel J. Simmens, and Stuart M. Sotsky, 'Patient Alliance Treatment Expectancies, and Outcome: Further Analyses, From the National Institute of Mental Health Treatment, of Depression Collaborative Research Program', *Journal of Consulting and Clinical Psychology* 70, no. 4 (2002), 1051–55

Moerman, Daniel E., 'The Meaning Response: Thinking About Placebos', *Pain Practice* 6, no. 4 (2006): 233–36

—— and Wayne B. Jonas, 'Deconstructing the Placebo Effect and Finding the Meaning Response', *Annals of Internal Medicine* 136 (2002): 471–76

Moncrieff, Joanna, 'The Creation of the Concept of an Antidepressant: An Historical Analysis', *Social Science & Medicine* 66, no. 11 (2008a): 2346–55

——, *The Myth of the Chemical Cure*, Basingstoke: Palgrave Macmillan, 2008b

——, S. Wessely and R. Hardy, 'Active Placebos Versus Antidepressants for Depression (Review)', *The Cochrane Library*, no. 4 (2005)

Montgomery, Guy H. and Irving Kirsch, 'Classical Conditioning and the Placebo Effect', *Pain* 72, no. 1–2 (1997): 107–13

—— and Irving Kirsch, 'Mechanisms of Placebo Pain Reduction: An Empirical Investigation', *Psychological Science* 7, no. 3 (1996): 174–76

Moseley, J. Bruce, K. O'Malley, N. J. Petersen, T. J. Menke, B. A. Brody, D. H. Kuykendall, J. C. Hollingsworth, C. M. Ashton and Nelda P. Wray, 'A Controlled Trial of Arthroscopic Surgery for Osteoarthritis of the Knee', *New England Journal of Medicine* 347 (2002): 81–88

Myers, M. G., J. A. Calms and J. Singer, 'The Consent Form as a Possible Cause of Side Effects', *Clinical Pharmacology & Therapeutics* 42 (1987): 250–53

Nathan, Peter and Martin E. P. Seligman, 'Editors' Note', *Prevention & Treatment*, Article 0002a (1998); http://www.journals.apa.org/prevention/volume1/pre0010002a.html

NCAM, 'St John's Wort'; http://nccam.nih.gov/health/stjohnswort/ataglance.htm

NICE, 'Depression: Management of Depression in Primary and Secondary Care. Clinical Practice Guideline No. 23', National Institute for Clinical Excellence; www.nice.org.uk/page.aspx?o = 235213

Nitzan, Uriel and Pesach Lichtenberg, 'Questionnaire Survey on Use of Placebo', *British Medical Journal* 329 (2004): 944–46

Park, Lee C. and Lino Covi, 'Nonblind Placebo Trial', *Archives of General Psychiatry* 12, no. 4 (1965): 336–45

Pavlov, Ivan P., *Conditioned Reflexes*, translated by G. V. Anrep, London: Oxford University Press, 1927

——, 'Physiology of Digestion: Nobel Lecture, 12 December 1904'; http://nobelprize.org/nobel_prizes/medicine/laureates/1904/pavlov-lecture.html

Philipp, Michael, Ralf Kohnen and Karl O. Hiller, 'Hypericum Extract Versus Imipramine or Placebo in Patients with Moderate Depression: Randomised Multicentre Study of Treatment for Eight Weeks', *British Medical Journal* 319 (1999): 1534–39

Pinquart, M., P. M. Duberstein and J. M. Lyness, 'Effects of Psychotherapy and Other Behavioral Interventions on Clinically Depressed Older Adults: A Meta-Analysis', *Aging & Mental Health* 11, no. 6 (2007): 645–57

Pogge, R., 'The Toxic Placebo', *Medical Times* 91 (1963): 773–78

Posternak, Michael A., Mark Zimmerman, Gabor I. Keitner and Ivan W. Miller, 'A Reevaluation of the Exclusion Criteria Used in Antidepressant Efficacy Trials', *American Journal of Psychiatry* 159 (2002): 191–200

Prator, Bettina C., 'Serotonin Syndrome', *Journal of Neuroscience Nursing* 38, no. 2 (2006): 102–05

Preskorn, Sheldon H., 'Tianeptine: A Facilitator of the Reuptake of Serotonin and Norepinephrine as an Antidepressant?', *Journal of Psychiatric Practice*, 10, no. 5 (2004): 323–30

Price, Donald D. and Howard I. Fields, 'The Contribution of Desire and Expectation to Placebo Analgesia: Implications for New Research Strategies', in *The Placebo Effect: An Interdisciplinary Exploration*, edited by Anne Harrington, Cambridge, MA: Harvard University Press, 1997; pp. 117–37

——, Leonard S. Milling, I. Kirsch, A. Duff, Guy H. Montgomery and S. S. Nicholls, 'An Analysis of Factors That Contribute to the Magnitude of Placebo Analgesia in an Experimental Paradigm', *Pain* 83, no. 2 (1999): 147–56

Quitkin, Frederic M., Patrick J. McGrath, Jonathan W. Stewart, Katja Ocepek-Welikson, Bonnie P. Taylor, Edward Nunes, Deborah Delivannides, Vito Agosti, Steven J. Donovan, Donald Ross, Eva Petkova and Donald F. Klein, 'Placebo Run-in Period in Studies of Depressive Disorders: Clinical, Heuristic and Research Implications', *British Journal of Psychiatry* 173 (1998): 242–48

Rabkin, J. G., J. S. Markowitz, J. W. Stewart, P. J. McGrath, W. Harrison, F. M. Quitkin and Donald F. Klein, 'How Blind Is Blind? Assessment of Patient and Doctor Medication Guesses in a Placebo-Controlled Trial of Imipramine and Phenelzine', *Psychiatry Research* 19 (1986): 75–86

Räikkönen, D., K. A. Matthews, J. D. Flory, J. F Owens and B. B. Gump, 'Effects of Optimism, Pessimism, and Trait Anxiety on Ambulatory Blood Pressure and Mood During Everyday Life', *Journal of Personality and Social Psychology* 76 (1999): 104–13

Reeves, Roy R., Mark E. Ladner, Roy H. Hart and Randy S. Burke, 'Nocebo Effects with Antidepressant Clinical Drug Trial Placebos', *General Hospital Psychiatry* 29 (2007): 275–77

Reiss, S. and R. J. McNally, 'The Expectancy Model of Fear', in *Theoretical Issues in Behavior Therapy*, edited by S. Reiss and Richard R. Bootzin, New York: Academic Press, 1985, pp. 107–21

Rescorla, Robert A., 'Pavlovian Conditioning: It's Not What You Think It Is', *American Psychologist* 43 (1988): 151–60

Robinson, Leslie A., Jeffrey S. Berman and Robert A. Neimeyer, 'Psychotherapy for the Treatment of Depression: A Comprehensive Review of Controlled Outcome Research', *Psychological Bulletin* 108, no. 1 (1990): 30–49

Robinson, Robert G., Ricardo E. Jorge, David J. Moser, Laura Acion, Ana Solodkin, Steven L. Small, Pasquale Fonzetti, Mark Hegel and Stephan Arndt, 'Escitalopram and Problem-Solving Therapy for Prevention of Poststroke Depression: A Randomized Controlled Trial', *Journal of the American Medical Association* 299, no. 20 (2008): 2391–400

Rogers, Carl, *On Becoming a Person: A Therapist's View of Psychotherapy*, London: Constable, 1961

Rothschild, Anthony J. and Carol A. Locke, 'Reexposure to Fluoxetine after Serious Suicide Attempts by Three Patients: The Role of Akathisia', *Journal of Clinical Psychiatry* 52, no. 12 (1991): 491–93

Ruhé, H. G., N. S. Mason and Aart H. Schene, 'Mood Is Indirectly Related to Serotonin, Norepinephrine and Dopamine Levels in Humans: A Meta-Analysis of Monoamine Depletion Studies', *Molecular Psychiatry* 12 (2007): 331–59

Rush, A. John, Madhukar H. Trivedi, Stephen R. Wisniewski, Andrew A. Nierenberg, Jonathan W. Stewart, Diane Warden, George Niederehe, Michael E. Thase, Philip W. Lavori, Barry D. Lebowitz, Patrick J. McGrath, Jerrold F. Rosenbaum, Harold A. Sackeim, David J. Kupfer, James Luther and Maurizio Fava, 'Acute and Longer-Term Outcomes in Depressed Outpatients Requiring One or Several Treatment Steps: A Star*D Report', *American Journal of Psychiatry* 163 (2006a): 1905–17

——, Madhukar H. Trivedi, Stephen R. Wisniewski, Jonathan W. Stewart, Andrew A. Nierenberg, Michael E. Thase, Louise Ritz, Melanie M. Biggs, Diane Warden, James F. Luther, Kathy Shores-Wilson, George Niederehe and Maurizio Fava, 'Bupropion-Sr, Sertraline, or Venlafaxine-Xr after Failure of SSRIs for Depression', *New England Journal of Medicine* 354 (2006b): 1231–42

Sarek, Milan, 'Evident Exception in Clinical Practice Not Sufficient to Break Traditional Hypothesis', *PLoS Medicine*, no. 2 (2006); www.plosmedicine.org

Scheier, M. F., K. A. Matthews, J. F. Owens, R. Schulz, M. W. Bridges, G. J. Magovern and C. S. Carver, 'Optimism and Rehospitalization after Coronary Artery Bypass Graft Surgery', *Archives of Internal Medicine* 159, no. 829–35 (1999)

Schildkraut, Joseph J., The Catecholamine Hypothesis of Affective Disorders: A Review of Supporting Evidence', *American Journal of Psychiatry* 122 (1965): 509–22

Schofield, William, *Psychotherapy: The Purchase of Friendship*, New Jersey: Prentice-Hall, 1964

Shapiro, Arthur K., 'A Contribution to a History of the Placebo Effect', *Behavioral Science* 5, no. 109–35 (1960)

—— and L. A. Morris, 'The Placebo Effect in Medical and Psychological Therapies', in *Handbook of Psychotherapy and Behavior Change*, edited by S. L. Garfield and A. E. Bergin, New York: Wiley, 1978

——, E. Struening and E. Shapiro, 'The Reliability and Validity of a Placebo Test', *Journal of Psychiatric Research* 15, no. 253–90 (1980)

Shepherd, Michael, 'Reserpine: Problems Associated with the Use of a So-Called "Tranquillizing Agent"', *Proceedings of the Royal Society of Medicine* 49, no. 10 (1956): 849–52

Sherman, Carl, 'Long-Term Side Effects Surface with SSRIs', *Clinical Psychiatry News* 26, no. 5 (1998): 1

Simpson, Scot H., Dean T. Eurich, Sumit R. Majumdar, Rajdeep S. Padwal, Ross T. Tsuyuki, Janice Varney and Jeffrey A. Johnson, 'A Meta-Analysis of the Association between Adherence to Drug Therapy and Mortality', *British Medical Journal* (2006)

Singh, Nalin A., Karen M. Clements and Maria A. Fiatarone Singh, 'The Efficacy of Exercise as a Long-Term Antidepressant in Elderly Subjects: A Randomized, Controlled Trial', *Journal of Gerontology* 56A, no. 8 (2001): M497–M504

Sloane, R. Bruce, Fred R. Staples, Allan H. Cristol, Neil J. Yorkston and Katherine Whipple, 'Patient Characteristics and Outcome in

Psychotherapy and Behavior Therapy', *Journal of Consulting and Clinical Psychology* 44, no. 3 (1976): 330–39

Smith, Nancy M., Mark R. Floyd, Forrest Scogin and Christine S. Jamison, 'Three-Year Follow-up of Bibliotherapy for Depression', *Journal of Consulting and Clinical Psychology* 65, no. 2 (1997): 324–27

Smith, Rebecca, 'Anti-Depressants "No Better Than Dummy Pills"', *Daily Telegraph*, 26 February 2008

Sneed, Joel R., Bret R. Rutherford, David Rindskopf, David T. Lane, Harold A. Sackeim and Steven P. Roose, 'Design Makes a Difference: A Meta-Analysis of Antidepressant Response Rates in Placebo-Controlled Versus Comparator Trials in Late-Life Depression', *American Journal of Geriatric Psychiatry* 16, no. 1 (2008): 65–73

Sodergren, Samantha C. and Michael E. Hyland, 'Expectancy and Asthma', in *How Expectancies Shape Experience*, edited by Irving Kirsch, Washington, DC: American Psychological Association, 1999, pp. 197–212

Sotsky, Stuart M., D. R. Glass, M. Tracie Shea, Paul A. Pilkonis, J. F. Collins, Irene Elkin, John T. Watkins, S. D. Imber, W. R. Leber and J. Moyer, 'Patient Predictors of Response to Psychotherapy and Pharmacotherapy: Findings in the NIMH Treatment of Depression Collaborative Research Program', *American Journal of Psychiatry* 148 (1991): 997–1008

Spiegel, David and Janine Giese-Davis, 'Depression and Cancer: Mechanisms and Disease Progression', *Biological Psychiatry* 54 (2003): 269–82

Sternberg, Esther M., 'Walter B. Cannon and "Voodoo Death": A Perspective from 60 Years On', *American Journal of Public Health* 92, no. 10 (2002): 1564–66

Stewart-Williams, Steve and John Podd, 'The Placebo Effect: Dissolving the Expectancy Versus Conditioning Debate', *Psychological Bulletin* 130, no. 2 (2004): 324–40

Stone, Marc B. and M. Lisa Jones, 'Clinical Review: Relationship between Antidepressant Drugs and Suicidality in Adults' (2006); http://www.fda.gov/ohrms/dockets/ac/06/briefing/2006–4272b1-01-FDA.pdf

Storms, Michael D. and Richard E. Nisbett, 'Insomnia and the Attribution Process', *Journal of Personality and Social Psychology* 16, no. 2 (1970): 319–28

Strawbridge, William J., Stéphane Deleger, Robert E. Roberts and George A. Kaplan, 'Physical Activity Reduces the Risk of Subsequent

Depression for Older Adults', *American Journal of Epidemiology* 156, no. 4 (2002): 328–34

Talbot, Margaret, 'The Placebo Prescription', *New York Times*, 9 January 2000

Taylor, Matthew J., Nick Freemantle, John R. Geddes and Zubin Bhagwagar, 'Early Onset of Selective Serotonin Reuptake Inhibitor Antidepressant Action: Systematic Review and Meta-Analysis', *Archives of General Psychiatry* 63 (2006): 1217–23

Teasdale, John D., 'Psychological Treatments for Depression: How Do They Work?', *Behaviour Research and Therapy* 23 (1985): 157–65

Thase, Michael E., 'Antidepressant Effects: The Suit May Be Small, but the Fabric Is Real', *Prevention & Treatment*, no. 32 (2002); http://www.journals.apa.org/prevention/volume5/pre0050032c.html

Tilburt, Jon C., Ezekiel J. Emanuel, Ted J. Kaptchuk, Farr A. Curlin and Franklin G. Miller, 'Prescribing "Placebo Treatments": Results of National Survey of US Internists and Rheumatologists', *British Medical Journal* 337 (2008): 1097–100

Traut, E. F. and E. W. Passarelli, 'Placebos in the Treatment of Rheumatoid Arthritis and Other Rheumatic Conditions', *Annals of the Rheumatic Diseases* 16 (1957): 18–22

Trivedi, Madhukar H., A. John Rush, Stephen R. Wisniewski, Andrew A. Nierenberg, Diane Warden, Louise Ritz, Grayson Norquist, Robert H. Howland, Barry Lebowitz, Patrick J. McGrath, Kathy Shores-Wilson, Melanie M. Biggs, G. K. Balasubramani, Maurizio Fava and STAR*D Study Team, 'Evaluation of Outcomes with Citalopram for Depression Using Measurement-Based Care in Star*D: Implications for Clinical Practice', *American Journal of Psychiatry* 163 (2006): 1–13

Turner, Erick H., Annette M. Matthews, Eftihia Linardatos, Robert A. Tell and Robert Rosenthal, 'Selective Publication of Antidepressant Trials and Its Influence on Apparent Efficacy', *New England Journal of Medicine* 358 (2008): 252–60

Tyrer, Peter, Patricia C. Oliver-Africano, Zed Ahmed, Nick Bouras, Sherva Cooray, Shoumitro Deb, Declan Murphy, Monica Hare, Michael Meade, Ben Reece, Kofi Kramo, Sabyasachi Bhaumik, David Harley, Adrienne Regan, David Thomas, Bharti Rao, Bernard North, Joseph Eliahoo, Shamshad Karatela, Anju Soni and Mike Crawford, 'Risperidone, Haloperidol, and Placebo in the Treatment of Aggressive Challenging Behaviour in Patients with Intellectual Disability: A Randomised Controlled Trial', *The Lancet* 371 (2008): 57–63

Ulrich, Roger S., 'View through a Window May Influence Recovery from Surgery', *Science* 224, no. 4647 (1984): 420–21

Uzbay, Tayfun I., 'Tianeptine: Potential Influences on Neuroplasticity and Novel Pharmacological Effects', *Progress in Neuro-Psychopharmacology & Biological Psychiatry* 32 (2008): 915–24

Vase, Lene, Joseph L. Riley III and Donald D. Price, 'A Comparison of Placebo Effects in Clinical Analgesic Trials Versus Studies of Placebo Analgesia', *Pain* 99 (2002): 443–52

Vedantam, Shankar, 'Fda Urged Withholding Data on Antidepressants; Makers Were Dissuaded from Labeling Drugs as Ineffective in Children', *Washington Post*, 10 September 2004

Voudouris, N. J., C. L. Peck and G. Coleman, 'Conditioned Placebo Responses', *Journal of Personality and Social Psychology* 48 (1985): 47–53

——, C. L. Peck and G. Coleman, 'Conditioned Response Models of Placebo Phenomena: Further Support', *Pain* 38 (1989): 109–16

——, C. L. Peck and G. Coleman, 'The Role of Conditioning and Verbal Expectancy in the Placebo Response', *Pain* 43 (1990): 121–28

Waber, Rebecca L., Baba Shiv, Ziv Carmon and Dan Ariely, 'Commercial Features of Placebo and Therapeutic Efficacy', *Journal of the American Medical Association* 299, no. 9 (2008): 1016–17

Wager, Tor D., 'The Neural Bases of Placebo Effects in Pain', *Current Directions in Psychological Science* 14, no. 4 (2005): 175–79

——, James K. Rilling, Edward E. Smith, Alex Sokolik, Kenneth L. Casey, Richard J. Davidson, Stephen M. Kosslyn, Robert M. Rose and Jonathan D. Cohen, 'Placebo-Induced Changes in fMRI in the Anticipation and Experience of Pain', *Science* 303, no. 20 (February 2004): 1162–67

——, David J. Scott and Jon-Kar Zubieta, 'Placebo Effects on Human Opioid Activity During Pain', *Proceedings of the National Academy of Sciences* 104, no. 26 (2007): 11056–61

Wagstaff, Antona J., Douglas Ormrod and Caroline M. Spencer, 'Tianeptine: A Review of Its Use in Depressive Disorders', *CNS Drugs* 15, no. 231–59 (2001)

Wampold, Bruce E., Takuya Minami, Thomas W. Baskin and Sandra Callen Tierney, 'A Meta-(Re)Analysis of the Effects of Cognitive Therapy Versus "Other Therapies" for Depression', *Journal of Affective Disorders* 68 (2002): 159–65

Warner, Christopher H., William Bobo, Carolynn Warner, Sara Reid and James Rachal, 'Antidepressant Discontinuation Syndrome', *American Family Physician* 74, no. 3 (2006): 449–56

Watson, Alison, Wael El-Deredy, Brent A. Vogt and Anthony K. P. Jones, 'Placebo Analgesia Is Not Due to Compliance or Habituation: EEG and Behavioural Evidence', *NeuroReport* 18, no. 8 (2007): 771–75

Weinberger, Joel and Andrew Eig, in *How Expectancies Shape Experience*, edited by Irving Kirsch, Washington, DC: American Psychological Association, 1999, pp. 357–82

Werner, Rachel, 'Losing the Point', *PLoS Medicine*, 28 February (2008); http://medicine.plosjournals.org/perlserv/?request-read-response&doi-10.1371/journal.pmed.0050045

Wilkinson, Richard and Kate Pickett, *The Spirit Level: Why More Equal Societies Almost Always Do Better*, London: Penguin Books, 2009

Williams, Jr, John W., Cynthia D. Mulrow, Elaine Chiquette, Polly Hitchcock Noel, Christine Aguilar and John Cornell, 'A Systematic Review of Newer Pharmacotherapies for Depression in Adults: Evidence Report Summary', *Annals of Internal Medicine* 132 (2000): 743–56

Wolf, S., 'Effects of Suggestion and Conditioning on the Action of Chemical Agents in Human Subjects – the Pharmacology of Placebos', *Journal of Clinical Investigation* 29 (1950): 100–09

——, C. R. Doering, M. L. Clark and J. A. Hagans, 'Chance Distribution and the Placebo "Reactor"', *Journal of Laboratory and Clinical Medicine* 49 (1957): 837–41

Wray, Nelda P., J. Bruce Moseley and K. O'Malley, 'Arthroscopic Surgery for Osteoarthritis of the Knee [Letter]', *New England Journal of Medicine* 234 (2002): 1718–19

Yang, H. and W. Lin, 'Effects of Positive and Negative Emotion on Neuroendocrine and Immunity', *Psychological Science* 28, no. 4 (2005): 926–28

Zahl, Per-Henrik, Jan Mæhlen and H. Gilbert Welch, 'The Natural History of Invasive Breast Cancers Detected by Screening Mammography', *Annals of Internal Medicine* 168, no. 21 (2008): 2311–16

Index